Applied Geophysics: A Contemporary View

Ndikum Eric Ndoh

DEDICATION

I heartily dedicate this book to my sweet wife Ndikum Henrietta and to our lovely kids Chani, Yadest, Jevilov & Yadav.

Foreword

In an era defined by critical global challenges—from ensuring sustainable energy and water resources to mitigating natural hazards and understanding our planet's deep processes—the field of geophysics has never been more vital. The Earth, our complex home, reveals its secrets not through direct observation alone, but through the subtle language of its physical fields and responses to applied energy. Interpreting this language is the art and science of geophysics.

This timely manuscript, "Applied Geophysics: A Contemporary View," offers a comprehensive and refreshingly accessible journey through the essential principles and cutting-edge applications that define modern geophysics. What sets this work apart is its admirable balance: it meticulously grounds the reader in the fundamental physics and historical evolution of geophysical methods, then boldly extends into the sophisticated technologies and multidisciplinary approaches that are shaping the discipline today.

From the foundational concepts of seismic wave propagation and potential fields to the revolutionary insights offered by satellite geodesy, distributed acoustic sensing (DAS), and the burgeoning role of machine learning, the authors provide a clear and insightful narrative. The dedicated chapters on mining and petroleum exploration are particularly commendable, showcasing the practical utility of these techniques in resource discovery, characterization, and responsible management. Furthermore, the explicit discussion of the societal impact and the field's increasing relevance to climate change, natural hazard mitigation, and critical mineral exploration truly highlights geophysics as a discipline at the forefront of global sustainability.

This book is more than a textbook; it is an invaluable guide for students entering the field, a practical reference for professionals seeking to update their knowledge, and an enlightening read for anyone curious about how we "see" the unseen beneath our feet. The clarity of explanation, coupled with a pragmatic focus on real-world applications and future trends, ensures that readers will not only grasp the concepts but also appreciate the dynamic and evolving nature of applied geophysics.

It is with great pleasure that I recommend "Applied Geophysics: A Contemporary View." It is poised to become a standard reference for the next generation of geoscientists who will undoubtedly contribute to a deeper understanding and more sustainable stewardship of our planet.

Prof. Koumetio Fidèle (Professor of Geophysics, University of Dschang) 30/07/2025

Preface

The Earth beneath our feet is a dynamic and intricate system, holding secrets vital to our very existence—from the resources that fuel our societies to the geological processes that shape our environment and pose hazards. For centuries, humanity has sought to understand these subsurface mysteries. In the modern era, the field of geophysics has emerged as the indispensable key to unlocking these hidden complexities, allowing us to "see" into the Earth without direct excavation. "Applied Geophysics: A Contemporary View" is born from a desire to provide a comprehensive, accessible, and forward-looking text for students, researchers, and professionals navigating this exciting and ever-evolving discipline. Our primary objective is to bridge the gap between fundamental geophysical principles and their practical applications in solving real-world problems. We believe that a deep appreciation for the underlying physics is paramount to effectively wielding these powerful tools in diverse scenarios, whether it be for resource exploration, environmental stewardship, or hazard assessment.

This manuscript endeavors to offer a holistic perspective on applied geophysics. We begin by grounding the reader in the basic tenets of general and applied geophysics, categorizing methods by their passive or active nature, and introducing the crucial concept of geophysical anomalies. A historical overview traces the remarkable journey of the field, from ancient observations to the transformative 20th-century breakthroughs and the current era of advanced computational power and global data integration.

The core of the book delves into the components of applied geophysics, systematically detailing both traditional and modern methods. We explore the principles, instrumentation, data acquisition techniques, and diverse applications of seismic, magnetic, gravity, electrical, electromagnetic, radiometric, and borehole geophysical methods. Crucially, dedicated chapters on mining and petroleum exploration highlight how these methods are specifically tailored and integrated to address the unique challenges and opportunities within these vital industries. We emphasize the synergistic value of multi-method surveys and the transformative impact of computational geophysics, machine learning, and advanced sensor technologies.

Perhaps most importantly, this book looks to the future. We extensively discuss the emerging trends, ongoing challenges, and the increasing societal relevance of applied geophysics in areas such as carbon capture and storage (CCUS), geothermal energy, critical mineral exploration, and natural hazard mitigation. We aim to equip the next generation of geoscientists not only with the technical knowledge but also with an understanding of their crucial role in addressing global environmental and resource challenges.

We have strived for clarity in explanation, ensuring that complex concepts are presented in an understandable manner, complemented by illustrative figures and real-world case studies. Our hope is that this book will serve as a foundational resource, inspiring readers to delve deeper into the fascinating world of applied geophysics and to contribute to its continued innovation and impact.

TABLE OF CONTENTS

ACKNOWLEDGMENTS

I greatly acknowledge USSGS for the images used from their image gallery.

CHAPTER 1 INTRODUCTION

1.1 General Geophysics

Definition: General geophysics is the study of the physical properties and processes of the Earth and its atmosphere, using quantitative methods (Lowrie, 2007).

Characteristics: General geophysics is a broad and interdisciplinary field that encompasses a wide range of topics, including:

- Seismology: the study of seismic waves and earthquakes (Kearey and Brooks, 2002)
- Gravity and geodesy: the study of the Earth's gravity field and its variations (Telford et al., 1990)
- Magnetics: the study of the Earth's magnetic field and its variations (Telford et al., 1990)
- Electromagnetics: the study of the Earth's electrical and magnetic conductivity (Telford et al., 1990)
- Heat flow: the study of the Earth's internal heat and its flow to the surface (Turcotte and Schubert, 2002)
- Geodynamics: the study of the forces and processes that drive the Earth's tectonic activity (Turcotte and Schubert, 2002)

Domains of applications: General geophysics research is essential for understanding the Earth's interior and its evolution, as well as the natural hazards that we face. General geophysics research is also used to develop new technologies for mineral exploration, groundwater exploration, environmental remediation, and other applications.

1.2 Applied Geophysics

Definition: Applied geophysics is the use of geophysical methods to solve practical problems (Lowrie, 2007).

Characteristics: Applied geophysics is a diverse field that uses a wide range of geophysical methods to solve a wide range of problems. Some common applied geophysics methods include:

- Seismic refraction and reflection: used to image the subsurface and identify geological features such as faults and folds (Kearey and Brooks, 2002)
- Gravity and magnetics: used to map the subsurface and identify anomalies that may be associated with mineral deposits or groundwater aquifers (Kearey and Brooks, 2002)

- Electrical resistivity and electromagnetic conductivity: used to map the subsurface and identify anomalies that may be associated with groundwater aquifers, environmental contamination, or engineering hazards (Telford et al., 1990)

Domains of applications: Applied geophysics is used in a wide range of industries and applications, including:

- Mineral exploration: used to identify and map potential mineral deposits
- Groundwater exploration: used to identify and map groundwater aquifers
- Environmental remediation: used to map the extent of environmental contamination and to monitor the effectiveness of remediation efforts
- Engineering site investigations: used to assess the geotechnical properties of a site and to identify potential hazards such as faults and sinkholes
- Oil and gas exploration: used to identify and map potential oil and gas reservoirs

1.3 Passive Methods of Geophysics

Passive geophysical methods measure the natural physical fields of the Earth, such as gravity, magnetic field, and seismic waves. These fields are generated by natural processes, such as the rotation of the Earth, the presence of iron-rich minerals in the crust, and the release of energy from earthquakes.

Examples of passive geophysical methods:

- Gravimetry: measures the Earth's gravity field
- Magnetometry: measures the Earth's magnetic field
- Seismology: measures seismic waves

Characteristics of passive geophysical methods:

- Passive geophysical methods are non-invasive, meaning that they do not require any energy to be added to the Earth system.
- Passive geophysical methods can be used to study a wide range of geological features, from the crust to the core of the Earth.
- Passive geophysical methods are often used to map large areas quickly and efficiently.

Domains of applications of passive geophysical methods:

- Mineral exploration: to identify and map potential mineral deposits
- Groundwater exploration: to identify and map groundwater aquifers
- Geotechnical engineering: to assess the geotechnical properties of a site

- Environmental remediation: to map the extent of environmental contamination
- Earthquake hazard assessment: to identify and assess earthquake hazards

1.4 Active Methods of Geophysics

Active geophysical methods introduce an artificial source of energy into the Earth system and measure the response of the subsurface. The energy source can be seismic waves, electrical current, or electromagnetic radiation.

Examples of active geophysical methods:

- Seismic reflection: uses seismic waves to image the subsurface
- Electrical resistivity tomography: uses electrical current to image the subsurface
- Ground penetrating radar (GPR): uses electromagnetic radiation to image the subsurface

Characteristics of active geophysical methods:

- Active geophysical methods can provide high-resolution images of the subsurface.
- Active geophysical methods can be used to study a wide range of geological features, from the shallow subsurface to the deep crust.
- Active geophysical methods are often more expensive and time-consuming to use than passive geophysical methods.

Domains of applications of active geophysical methods:

- Mineral exploration: to identify and map potential mineral deposits
- Groundwater exploration: to identify and map groundwater aquifers
- Geotechnical engineering: to assess the geotechnical properties of a site
- Environmental remediation: to map the extent of environmental contamination
- Archaeological research: to locate and map buried archaeological features

1.5 Geophysical Anomaly

A geophysical anomaly is a variation in the Earth's physical fields that cannot be explained by the normal background field. Geophysical anomalies can be caused by a variety of factors, such as the presence of mineral deposits, groundwater aquifers, or geological faults.

Characteristics of geophysical anomalies:

- Geophysical anomalies can be positive or negative, meaning that the measured value is higher or lower than the expected background value.
- Geophysical anomalies can be localized or extensive.
- Geophysical anomalies can be caused by a variety of different geological features.

Domains of applications of geophysical anomalies:

- Mineral exploration: to identify potential mineral deposits
- Groundwater exploration: to identify potential groundwater aquifers
- Geotechnical engineering: to identify potential geological hazards
- Environmental remediation: to identify the extent of environmental contamination
- Archaeological research: to locate buried archaeological features

Measurement of the Anomaly

Geophysical anomalies are measured using a variety of instruments, depending on the specific geophysical method being used. For example, gravity anomalies are measured using gravimeters, magnetic anomalies are measured using magnetometers, and seismic anomalies are measured using geophones.

Examples of instruments for geophysical measurements:

- Gravimeter: measures the Earth's gravity field
- Magnetometer: measures the Earth's magnetic field
- Geophone: measures seismic waves
- Electrical resistivity meter: measures the electrical resistivity of the subsurface
- Ground penetrating radar: measures the electromagnetic response of the subsurface

1.6 Functioning principles of instruments for geophysical measurements

Here presented are the fundamental physical principles that underpin the instruments used to acquire geophysical data. Understanding these principles is crucial for appreciating the strengths and limitations of different measurement techniques and for interpreting the resulting data effectively.

Gravity Meters

Gravity meters, also known as gravimeters, are precision instruments designed to measure the local gravitational field of the Earth. They detect minute variations in gravitational acceleration caused by differences in subsurface density.

Figure 1 A gravimeter which was used to measure the force of gravity on Mauna Kea. These measurements were used to calibrate the instruments so they could precisely monitor changes in gravity from magma accumulation at Mauna Loa (background). USGS photograph taken on December 10, 2020. (Courtesy USGS Multimedia Gallery)

- **Functioning Principle:** Most modern land gravity meters are based on the principle of a spring balance system. A mass is suspended by a sensitive spring, and the extension of the spring is proportional to the gravitational force acting on the mass. Changes in the gravitational field cause minute changes in the spring's extension, which are precisely measured using electronic or optical techniques. Relative gravity meters measure the difference in gravity between two or more points, while absolute gravity meters determine the absolute value of gravitational acceleration at a specific location (Telford et al., 1990). Superconducting gravimeters, a more advanced type, utilize the levitation of a superconducting sphere in a magnetic field to achieve extremely high sensitivity (Goodkind, 1999).

- **Characteristics:**
  - **Sensitivity:** High sensitivity to small variations in gravity (microgals or even nanogals for superconducting gravimeters).
  - **Accuracy:** Achieving high accuracy requires careful calibration and corrections for factors like Earth tides, instrument drift, and terrain effects.
  - **Portability:** Land gravity meters are designed to be portable for field surveys, while borehole gravimeters are specialized for measurements within boreholes. Marine and airborne gravimeters require sophisticated stabilization platforms.
  - **Measurement Rate:** Relatively slow measurement rate compared to some other geophysical methods.
- **Domains of Applications:**
  - **Resource Exploration:** Identifying subsurface density contrasts associated with mineral deposits, oil and gas reservoirs, and geothermal resources (Reynolds, 2011).
  - **Hydrogeology:** Mapping aquifers and investigating groundwater storage.
  - **Geodynamics:** Studying crustal deformation, tectonic processes, and post-glacial rebound.
  - **Engineering Geophysics:** Detecting cavities, buried structures, and assessing ground stability.
  - **Archaeology:** Locating buried archaeological features.

Magnetometers

Magnetometers are instruments used to measure the strength and/or direction of magnetic fields, particularly the Earth's magnetic field and local anomalies caused by magnetized rocks or artificial objects.

- **Functioning Principle:** Various types of magnetometers operate based on different physical principles:
  - **Fluxgate Magnetometers:** These measure the change in permeability of a highly permeable core when subjected to an external magnetic field. An alternating current

is passed through a primary coil, and the induced voltage in a secondary coil is proportional to the external magnetic field strength (Parasnis, 1997).

Figure 2: A Piper Navajo airplane with auxiliary wingtip pods and tail stinger magnetometers used in studying the rocks within the St. Francois Mountains in southeast Missouri. (Courtesy USGS Multimedia Gallery)

- **Proton Precession Magnetometers (PPM):** These utilize the precession of protons (hydrogen nuclei) in a fluid-filled sensor when subjected to a strong, temporarily applied magnetic field. The frequency of the precessing protons is directly proportional to the magnitude of the Earth's magnetic field (Breiner, 1973).
- **Optically Pumped Magnetometers (e.g., Cesium, Potassium):** These are highly sensitive instruments that exploit the quantum mechanical properties of alkali metals. Light of a specific wavelength is used to pump electrons to a higher energy level. The absorption of light changes when the atoms are subjected to a magnetic field, and this change is measured to determine the field strength (Telford et al., 1990).
- **SQUID Magnetometers (Superconducting Quantum Interference Devices):** These are extremely sensitive magnetometers that utilize the quantum mechanical phenomenon of superconductivity to measure minute changes in magnetic flux. They are often used in specialized research applications.

- **Characteristics:**
    - **Sensitivity:** Varies greatly depending on the type, ranging from nT (nanotesla) for fluxgate and PPM to fT (femtotesla) for SQUIDs.
    - **Accuracy:** Influenced by factors such as instrument calibration, diurnal variations of the Earth's magnetic field, and magnetic noise.
    - **Measurement Rate:** Can range from relatively slow for some PPMs to very fast for optically pumped and fluxgate magnetometers.
    - **Applications:** Airborne, ground, and marine surveys are common.
- **Domains of Applications:**
    - **Mineral Exploration:** Locating magnetic ore bodies (e.g., iron, nickel).
    - **Archaeology:** Detecting buried magnetic features like kilns, hearths, and foundations.
    - **Environmental Geophysics:** Mapping buried metallic objects, landfills, and contaminated sites.
    - **Geomagnetism:** Studying the Earth's magnetic field, its variations, and paleomagnetism.
    - **Security and Military Applications:** Detecting buried or submerged metallic objects.

Seismic Instruments (Geophones and Hydrophones)

Seismic instruments are used to detect and record ground vibrations (seismic waves) generated by natural sources (earthquakes) or artificial sources (explosions, vibrators). Geophones are designed for land-based measurements, while hydrophones are used in aquatic environments.

- **Functioning Principle:**
    - **Geophones:** Most common geophones are based on the principle of electromagnetic induction. A mass suspended by a spring moves relative to a coil when the ground vibrates. This relative motion induces a voltage in the coil, which is proportional to the velocity of the ground motion (Sheriff & Geldart, 1995). Different types of geophones are designed to be sensitive to vertical or horizontal ground motion. Accelerometers, which measure ground acceleration, are also used in some seismic applications.

- **Hydrophones:** These are pressure-sensitive transducers that convert pressure variations in water caused by seismic waves into electrical signals. Piezoelectric crystals are commonly used in hydrophones; when subjected to pressure changes, they generate an electrical charge proportional to the pressure (Kinsler et al., 2000).

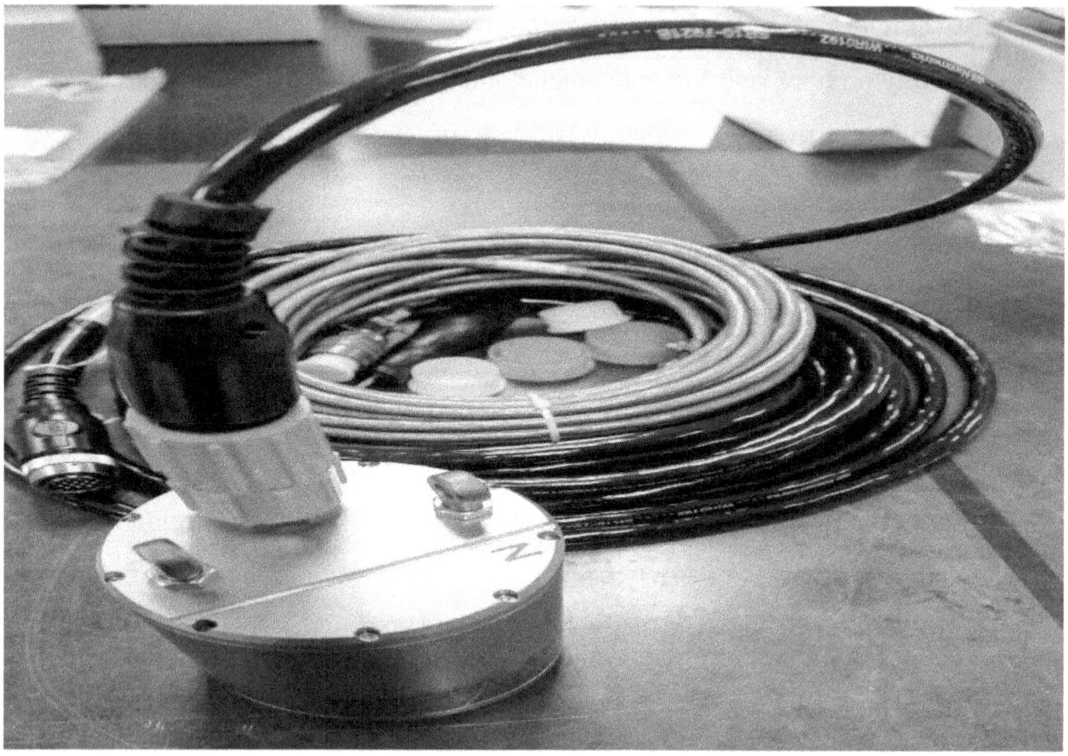

Figure 3: Seismometers (instruments for recording earthquakes) which were tested and fitted at the USGS Cascades Volcano Observatory before going out into the field. (Courtesy USGS Multimedia Gallery)

- **Characteristics:**
  - **Sensitivity:** Dependent on the design and intended frequency range.
  - **Frequency Response:** Geophones and hydrophones are designed to operate within specific frequency bands relevant to the seismic source and target.
  - **Damping:** Critical for controlling the instrument's response and preventing oscillations.
  - **Coupling:** Good coupling with the ground (for geophones) or water (for hydrophones) is essential for accurate measurements.

- **Domains of Applications:**
  - **Oil and Gas Exploration:** Mapping subsurface geological structures to identify potential hydrocarbon reservoirs (seismic reflection and refraction surveys).
  - **Earthquake Seismology:** Studying earthquakes, their sources, and the Earth's interior.
  - **Engineering Seismology:** Assessing site response to seismic shaking and evaluating structural integrity.
  - **Mining Geophysics:** Detecting subsurface cavities and monitoring ground stability.
  - **Marine Geophysics:** Studying sub-seafloor structures and processes.

Electrical and Electromagnetic (EM) Instruments

These instruments measure the electrical properties (e.g., resistivity, conductivity, induced polarization) and electromagnetic responses of the subsurface to infer geological structures, groundwater distribution, and the presence of conductive materials.

- **Functioning Principle:** A wide range of techniques and instruments fall under this category:
  - **DC Resistivity Methods:** These involve injecting direct current (or low-frequency AC) into the ground through electrodes and measuring the resulting potential difference at other electrodes. The apparent resistivity of the subsurface is then calculated based on the current and voltage measurements and the electrode geometry (Keller & Frischknecht, 1966). Different electrode arrays (e.g., Wenner, Schlumberger, dipole-dipole) are used for different investigation depths and target geometries.

Figure 4 USGS hydrologist conducting a two-dimensional (2D) resistivity survey to investigate and characterize the shallow subsurface as part of an applied research effort by the USGS Office of Groundwater Branch of Geophysics in 2006. (Courtesy USGS Multimedia Gallery)

- **Induced Polarization (IP):** This technique measures the ability of the ground to store electrical charge when subjected to a current flow and then slowly release it after the current is turned off. This phenomenon is often associated with the presence of polarizable materials like metallic sulfides or clay minerals (Telford et al., 1990). IP is typically measured in the time domain (measuring the decay of voltage after current shut-off) or the frequency domain (measuring the change in resistivity with frequency).
- **Electromagnetic (EM) Methods:** These techniques utilize the principles of electromagnetic induction. A time-varying magnetic field is generated by a transmitter coil, which induces eddy currents in conductive subsurface materials. These eddy currents, in turn, generate a secondary magnetic field that is detected by a receiver coil. The amplitude, phase, and frequency dependence of the secondary field provide information about the conductivity distribution in the

subsurface (Nabighian, 1987). EM methods can be ground-based (e.g., VLF, Slingram, transient EM), airborne (e.g., frequency-domain EM, time-domain EM), or borehole-based.

Figure 5 USGS hydrologist conducting a broadband electromagnetic survey in New Orleans, Louisiana. The research was conducted by the USGS Office of Groundwater Branch of Geophysics in 2006. (Courtesy USGS Multimedia Gallery)

  - **Magnetotellurics (MT):** This passive EM method utilizes naturally occurring variations in the Earth's magnetic and electric fields to investigate the subsurface conductivity structure at great depths.
- **Characteristics:**
  - **Depth of Investigation:** Varies significantly depending on the method and instrument configuration, ranging from a few meters to hundreds of kilometers.
  - **Sensitivity:** Dependent on the conductivity contrasts in the subsurface and the noise levels.

    - **Resolution:** Influenced by the frequency of the applied field (for EM methods) and the electrode spacing (for DC resistivity and IP).
    - **Portability:** Ranges from small, portable instruments for shallow surveys to large, truck-mounted systems for deep investigations.
- **Domains of Applications:**
    - **Groundwater Exploration:** Locating aquifers and mapping saltwater intrusion.
    - **Mineral Exploration:** Detecting conductive ore bodies.
    - **Environmental Geophysics:** Mapping contamination plumes, buried waste, and landfills.
    - **Geothermal Exploration:** Identifying zones of high conductivity associated with geothermal resources.
    - **Engineering Geophysics:** Investigating soil and rock properties, detecting cavities, and assessing corrosion potential.
    - **Archaeology:** Locating buried conductive features.

Radiometric Instruments (Gamma-Ray Spectrometers)

Radiometric instruments, particularly gamma-ray spectrometers, measure the natural radioactivity of rocks and soils by detecting the gamma rays emitted during the radioactive decay of elements such as potassium (K), uranium (U), and thorium (Th).

- **Functioning Principle:** Gamma-ray spectrometers typically use scintillation detectors or semiconductor detectors to measure the energy and intensity of incoming gamma rays. Different radioactive elements emit gamma rays with characteristic energy spectra, allowing for the identification and quantification of their concentrations in the ground (IAEA, 2003).
- **Characteristics:**
    - **Sensitivity:** Depends on the size and efficiency of the detector and the counting time.
    - **Spatial Resolution:** Influenced by the sensor footprint and the measurement height (for airborne surveys).
    - **Measurement Rate:** Can be relatively fast, allowing for efficient airborne surveys.

- **Penetration Depth:** Gamma rays have limited penetration in the ground, so the measurements primarily reflect the radioactivity of the near-surface materials (top few tens of centimeters).

- **Domains of Applications:**
  - **Geological Mapping:** Differentiating lithologies based on their distinct radiometric signatures.
  - **Mineral Exploration:** Identifying deposits of uranium, thorium, and potassium, as well as pathfinder elements associated with other types of mineralization.
  - **Environmental Monitoring:** Mapping areas with elevated levels of natural or artificial radioactivity.
  - **Agricultural Studies:** Assessing soil potassium content.
  - **Land Use Planning:** Providing information on the natural radiation background.

Remote Sensing Instruments (e.g., Multispectral and Hyperspectral Scanners, Thermal Imagers, LiDAR)

Remote sensing instruments acquire information about the Earth's surface from a distance, typically from aircraft or satellites, by measuring the electromagnetic radiation reflected or emitted by the surface.

- **Functioning Principle:** These instruments operate across different parts of the electromagnetic spectrum:
  - **Multispectral and Hyperspectral Scanners:** These sensors measure the intensity of reflected or emitted radiation in multiple discrete (multispectral) or continuous (hyperspectral) spectral bands. The spectral signature of different materials can be used to identify and map them (Lillesand et al., 2015).
  - **Thermal Imagers:** These sensors detect the thermal infrared radiation emitted by the Earth's surface, providing information about surface temperature variations.
  - **Light Detection and Ranging (LiDAR):** This active remote sensing technique uses a laser to measure the distance to the Earth's surface. By emitting rapid pulses of laser light and measuring the time it takes for the reflected light to return, LiDAR can create high-resolution 3D models of the terrain and vegetation canopy (Shan & Toth, 2009).

Figure 6: USGS Geographer Josh Logan sets up the lidar scanner near Capitola before the December 11, 2014 "Super Soaker" storm. (Courtesy USGS Multimedia Gallery)

- **Characteristics:**
  - **Spatial Resolution:** Varies greatly depending on the sensor and platform, ranging from centimeters to kilometers.
  - **Spectral Resolution:** The number and width of the spectral bands. Hyperspectral sensors provide much finer spectral resolution than multispectral sensors.
  - **Temporal Resolution:** The frequency with which data is acquired over a particular area.
  - **Coverage:** Can range from small local areas to global coverage.
- **Domains of Applications:**
  - **Geological Mapping:** Identifying lithological units, structural features, and mineral alteration zones.
  - **Environmental Monitoring:** Mapping land cover change, deforestation, water quality, and natural hazards.

  - **Hydrology:** Studying surface water bodies, snow cover, and soil moisture.
  - **Agriculture:** Monitoring crop health and yield.
  - **Urban Planning:** Creating detailed maps of urban areas and infrastructure.
  - **Archaeology:** Identifying potential archaeological sites through vegetation anomalies or subtle topographic features.
  - **Natural Hazard Assessment:** Mapping landslides, floods, and volcanic activity.

1.7 Data Acquisition

Geophysical data acquisition is the process of collecting data from geophysical instruments. The specific data acquisition process will vary depending on the type of geophysical instrument being used.

General steps in geophysical data acquisition:

1. **Set up the instrument:** The geophysical instrument should be set up in the desired location and calibrated.
2. **Collect the data:** The data should be collected according to the manufacturer's instructions.
3. **Store the data:** The data should be stored in a digital format for further processing and interpretation.

1.8 Data Processing

Raw geophysical data often contain noise and artifacts that obscure the signals of interest. Data processing techniques are employed to enhance the signal-to-noise ratio and prepare the data for interpretation.

Data processing is therefore the application of various techniques to transform raw geophysical data into a more interpretable form.

- **Characteristic elements**:
  - Noise Reduction: Filtering, stacking, and other techniques to remove unwanted signals.
  - Correction: Applying corrections for instrument drift, topography, and other systematic errors.
  - Enhancement: Applying techniques to improve the visibility of relevant features.
  - Imaging: Transforming the data into a spatial representation of the subsurface.

- **Domains of Applications**: All geophysical methods require data processing. For example, seismic data processing, gravity data processing, etc.

1.9 Data Interpretation

Geophysical data interpretation is the process of analyzing geophysical data to identify and characterize geological features.

Geophysical data interpretation is a complex process that requires a deep understanding of the physical principles underlying geophysical methods, as well as the geological context of the study area. It is often iterative, with the interpreter going back and forth between the data and their geological model to refine their understanding of the subsurface.

Geophysical data interpretation is used in a wide range of applications, including mineral exploration, groundwater exploration, geotechnical engineering, environmental remediation, and archaeological research.

Domains of applications:

- Mineral exploration: to identify and characterize potential mineral deposits.
- Groundwater exploration: to identify and characterize groundwater aquifers.
- Geotechnical engineering: to assess the geotechnical properties of a site.
- Environmental remediation: to map the extent of environmental contamination and to monitor the effectiveness of remediation efforts.
- Archaeological research: to locate and characterize buried archaeological features.

1.10 Forward and Inverse Modelling

Forward modelling is the process of predicting the response of a geophysical instrument to a known geological model. Inverse modelling is the process of inferring the geological model from the response of a geophysical instrument.

The following should be noted:

- Forward and inverse modelling are both important tools for geophysical data interpretation.
- Forward modelling can be used to generate synthetic data that can be compared to real data to validate the interpreter's geological model.
- Inverse modelling can be used to generate a geological model that fits the real data.

Domains of applications:

- Mineral exploration: to predict the response of geophysical instruments to potential mineral deposits and to infer the geological structure of the subsurface.

- Groundwater exploration: to predict the response of geophysical instruments to groundwater aquifers and to infer the geological structure of the subsurface.
- Geotechnical engineering: to predict the response of geophysical instruments to geotechnical hazards such as faults and sinkholes and to infer the geotechnical properties of a site.
- Environmental remediation: to predict the response of geophysical instruments to environmental contamination and to infer the extent of the contamination.
- Archaeological research: to predict the response of geophysical instruments to buried archaeological features and to infer the location and nature of the features.

1.11 Non-Uniqueness and Ambiguity

Non-uniqueness and ambiguity are two important concepts in geophysical data interpretation. Non-uniqueness refers to the fact that there may be multiple geological models that can fit the same geophysical data. Ambiguity refers to the difficulty in distinguishing between different geological models that fit the same geophysical data.

The following can be noted about them:

- Non-uniqueness and ambiguity are inherent in geophysical data interpretation because geophysical data can be affected by a variety of factors, including the geological complexity of the subsurface, the type of geophysical method being used, and the quality of the data.
- Non-uniqueness and ambiguity can be mitigated by using multiple geophysical methods, collecting high-quality data, and incorporating prior geological knowledge into the interpretation process.

Domains of applications:

- Mineral exploration: to reduce the risk of drilling into a barren hole.
- Groundwater exploration: to ensure that the location of a well will intersect a productive aquifer.
- Geotechnical engineering: to avoid constructing a structure on a hazardous site.
- Environmental remediation: to ensure that the extent of environmental contamination is fully mapped.

- Archaeological research: to reduce the risk of damaging buried archaeological features during excavation.

Non-Uniqueness

Non-uniqueness in geophysical data interpretation is the possibility of multiple geological models explaining the same observed data. This is because the relationship between the physical properties of the subsurface and the response of geophysical instruments is complex and often ambiguous. It has the following characteristics:

- Non-uniqueness is a fundamental property of geophysical data interpretation and is not always avoidable.
- The degree of non-uniqueness depends on a number of factors, including the complexity of the subsurface geology, the type of geophysical method used, and the quality of the data.
- Non-uniqueness can be mitigated by using multiple geophysical methods, collecting high-quality data, and incorporating prior geological knowledge into the interpretation process.

Domains of applications:

Non-uniqueness is a concern in all areas of applied geophysics, but it is particularly important in applications where high-stakes decisions are being made, such as mineral exploration, groundwater exploration, and geotechnical engineering.

- **Mineral exploration:** Non-uniqueness can lead to the drilling of barren holes, which can be a significant financial loss.
- **Groundwater exploration:** Non-uniqueness can lead to the drilling of wells that do not intersect productive aquifers, which can lead to water shortages.
- **Geotechnical engineering:** Non-uniqueness can lead to the construction of structures on hazardous sites, which can result in safety failures and property damage.

Ambiguity

Ambiguity in geophysical data interpretation is the difficulty in distinguishing between different geological models that fit the same observed data. This is often caused by non-uniqueness, but it can also be caused by other factors, such as the limited resolution of geophysical methods.

It should be noted that:

- Ambiguity is a common problem in geophysical data interpretation, and it can be difficult to resolve.

- The degree of ambiguity depends on the same factors that control non-uniqueness, as well as the specific geological models being considered.
- Ambiguity can be mitigated by using multiple geophysical methods, collecting high-quality data, incorporating prior geological knowledge into the interpretation process, and using quantitative methods to evaluate the different models.

Domains of applications:

Ambiguity is a concern in all areas of applied geophysics, but it is particularly important in applications where the consequences of misinterpreting the data are high, such as mineral exploration, groundwater exploration, and geotechnical engineering.

- **Mineral exploration:** Ambiguity can lead to the drilling of barren holes or the selection of the wrong mineral deposit for development.
- **Groundwater exploration:** Ambiguity can lead to the drilling of wells that do not intersect productive aquifers or the overexploitation of groundwater resources.
- **Geotechnical engineering:** Ambiguity can lead to the construction of structures on hazardous sites or the design of structures that are not strong enough to withstand the subsurface conditions.

CHAPTER 2: HISTORY OF GEOPHYSICS

2.1 Early History

This section will explore the nascent stages of geophysical thought and practice, tracing its roots from ancient observations to the foundational scientific principles that laid the groundwork for modern geophysics.

- **Ancient Observations and Proto-Geophysical Concepts:**

Geophysics, as a formalized scientific discipline, is a relatively modern invention. However, its roots stretch back to ancient civilizations, where curious minds observed natural phenomena and implicitly, or sometimes explicitly, grappled with concepts that would later form the bedrock of geophysical understanding.

Early Understanding of Magnetism: One of the earliest and most fascinating proto-geophysical concepts revolved around magnetism. The mysterious allure of **lodestone**, naturally magnetized iron ore, captivated ancient peoples. Its ability to attract iron was noted by various cultures independently. Perhaps the most significant early advancement came from China, where the **compass** was developed. This remarkable invention, initially used for geomancy and later for navigation, demonstrated a practical understanding of Earth's magnetic field, long before its true nature was comprehended. In the Western world, **Thales of Miletus** (c. 624/623 – c. 548/545 BC), one of the earliest Greek philosophers, is credited with observing and pondering the properties of magnetism and static electricity (e.g., in his observations of amber). While his explanations were philosophical rather than scientific in the modern sense, his recognition of these forces marked an important step towards understanding the physical world.

Rudimentary Understanding of Earthquakes and Volcanic Activity: Earthquakes and volcanic eruptions, with their raw power and destructive potential, naturally demanded explanations from ancient civilizations. Across diverse cultures, myths and early theories emerged to account for these terrifying events. While often attributed to divine wrath or mythical creatures, these explanations represented early attempts to make sense of geological processes. In the Greek tradition, **Aristotle's "Meteorology"** (c. 350 BC) offered a more systematic, albeit largely incorrect by modern standards, explanation for earthquakes. He proposed that subterranean winds, trapped within the Earth, caused tremors when they tried to escape. Despite its inaccuracies, Aristotle's work was highly influential for centuries and represented a significant early effort to provide a natural, rather than purely supernatural, explanation for geological phenomena. Similarly, observations of volcanic eruptions, particularly in regions like the Mediterranean, led to early ideas about underground fires and the release of Earth's internal heat.

Ancient Mining Practices and Implicit Geological Understanding: Long before the term "geophysics" existed, ancient mining practices inherently relied on a pragmatic, though unformalized, understanding of geological principles. Miners instinctively knew that certain rocks and geological formations were associated with valuable mineral deposits. They developed techniques for prospecting, identifying veins, and excavating ore, which demonstrated an implicit grasp of rock types, geological structures, and the distribution of resources within the Earth. The ability to locate and extract metals like copper, tin, gold, and silver for tools, weapons, and ornamentation, often from challenging geological environments, is testament to their sophisticated, albeit empirical, geological knowledge. These practices, driven by necessity and observation, laid a practical foundation for the later, more theoretical development of geological and geophysical sciences.

- **The Renaissance and the Scientific Revolution:**

The Renaissance and the ensuing Scientific Revolution marked a pivotal shift in humanity's understanding of the natural world, moving from speculative philosophy towards systematic observation, experimentation, and mathematical formulation. This era laid many of the crucial intellectual and practical foundations for what would later become modern geophysics.

Gravity: From Falling Bodies to Universal Laws: The concept of gravity underwent a radical transformation during this period. **Galileo Galilei's** (1564-1642) groundbreaking experiments on falling bodies, often associated with the Leaning Tower of Pisa, challenged Aristotelian dogma. His meticulous observations demonstrated that objects fall at the same rate regardless of their mass, a fundamental insight into the nature of gravitational acceleration. Building upon Galileo's work, **Isaac Newton** (1642-1727) formulated his groundbreaking **Law of Universal Gravitation**, published in his *Principia Mathematica* (1687). This law, which states that every particle in the universe attracts every other particle with a force directly proportional to the product of their masses and inversely proportional to the square of the distance between their centers, provided a complete theoretical framework for understanding the force that governs both celestial motions and phenomena on Earth. Crucially for geophysics, Newton's law provided the theoretical underpinning for **gravimetric surveys**, techniques that measure variations in the Earth's gravitational field to infer subsurface density variations, a cornerstone of modern geophysical exploration.

Magnetism: Earth as a Giant Magnet: The enigmatic force of magnetism also received profound attention. **William Gilbert's** (1544-1603) seminal work, *De Magnete, Magneticisque Corporibus, et Magno magnete Tellure* (On the Magnet and Magnetic Bodies, and on that Great Magnet the Earth), published in 1600, was a landmark achievement. Through meticulous experimentation, Gilbert demonstrated that the Earth itself behaves like a giant magnet, with magnetic poles. He differentiated between magnetism and static electricity, clarifying long-held misconceptions. His work marked a crucial step in understanding the Earth's magnetic field, moving beyond mere observation of lodestone to conceptualizing the planet as a dynamic magnetic entity. This understanding was vital for the later development of geomagnetism, a key branch of geophysics.

Early Seismology: Philosophical Stirrings: While instrumental seismology was still centuries away, the Renaissance and Scientific Revolution saw important philosophical discussions and early attempts to classify earthquakes. Thinkers began to move beyond purely supernatural explanations, seeking natural causes. Although instruments to precisely measure earthquake tremors were absent, observations of ground motion, associated sounds, and the extent of damage began to be systematically noted. **John Michell** (1724-1793), an English clergyman and natural philosopher, made particularly insightful contributions. As discussed by Oldroyd (1990), Michell proposed that earthquakes were caused by shifting masses of rock underground and, remarkably, suggested that they propagated as waves through the Earth. He also correctly deduced that the focus of an earthquake was often not directly beneath the epicenter. These visionary ideas, though lacking instrumental verification at the time, laid intellectual groundwork for the future science of seismology.

Mapping and Geodesy: Precision in Measurement: The drive for accuracy and precision during this era directly contributed to the development of **geodesy**, the scientific discipline of measuring and representing the Earth's shape, orientation, and gravity field. Early efforts in accurate surveying and mapping, spurred by navigation, commerce, and national defense, led to advancements in triangulation and the development of more precise instruments. The need to create accurate charts for exploration and trade highlighted the importance of understanding the Earth's curvature. These early mapping endeavors, though not explicitly "geophysical," provided the fundamental spatial framework and measurement techniques that are essential for all modern geophysical investigations, from seismic surveys to satellite-based observations of the Earth's surface.

2.2 The 20th Century

The 20th century marked a transformative era for geophysics, elevating it from a collection of nascent concepts to a robust and indispensable scientific discipline. This rapid evolution was fueled by a convergence of technological breakthroughs, the practical demands of industry and wartime, and ultimately, the revolutionary unifying theory of plate tectonics.

Emergence of Applied Geophysics: The early 20th century witnessed the birth of **applied geophysics**, driven largely by the insatiable global demand for energy resources. **Petroleum exploration** became a primary catalyst. The introduction of **seismic refraction and reflection methods**, pioneered by individuals like **Ludger Mintrop** with his early seismic surveys, revolutionized the search for oil and gas. The necessity of securing energy supplies, particularly during and after World War I, significantly accelerated the development and deployment of these techniques. Beyond oil, **mining geophysics** also saw significant advancements. Magnetic and gravimetric methods were increasingly applied to the efficient exploration for mineral deposits, alongside the nascent development of electrical and electromagnetic techniques.

Instrumental Development and Theoretical Advancements: The century was characterized by relentless innovation in geophysical instrumentation. In **seismology**, the development of increasingly sophisticated seismographs allowed for more precise recording and analysis of seismic waves. These advancements, coupled with meticulous seismological studies, began to unveil the Earth's enigmatic internal structure. Key discoveries included the **Mohorovičić discontinuity** (Moho), marking the crust-mantle boundary, and the **Gutenberg discontinuity**, delineating the core-mantle boundary. A crowning achievement was the work of **Inge Lehmann**, who in 1936, analyzing seismic waves, famously discovered the Earth's **inner core**, providing crucial insights into the planet's deep interior.

Gravity measurements saw a progression from rudimentary pendulum measurements to highly precise torsion balances and eventually to modern spring gravimeters. These improvements enabled detailed **gravity anomaly interpretations** for mapping subsurface geological structures. In **magnetism**, the development of more sensitive instruments like fluxgate magnetometers and proton precession magnetometers broadened their applications. Beyond traditional geological mapping, these advanced magnetometers found critical use in detecting subtle magnetic variations for purposes such as identifying **unexploded ordnance (UXO)**. Concurrently, **electrical and electromagnetic (EM) methods** diversified, with the development of various resistivity, induced polarization (IP), and EM techniques becoming essential tools for groundwater exploration, mineral prospecting, and environmental investigations.

The Plate Tectonics Revolution: The latter half of the 20th century was dominated by the **Plate Tectonics Revolution**, a paradigm shift that fundamentally transformed our understanding of Earth dynamics. Geophysical data played a crucial and undeniable role in solidifying this theory. **Seafloor spreading**, initially a bold hypothesis, found compelling evidence through **paleomagnetic studies**, most notably the work of **Vine and Matthews (1963)**, who demonstrated the symmetrical patterns of marine magnetic anomalies across mid-oceanic ridges. This provided direct proof of new oceanic crust being generated and spreading outwards. Furthermore, the global distribution of **seismicity patterns**, with earthquakes concentrated along narrow belts, provided strong evidence for the existence and movement of discrete **plate boundaries**. The ultimate acceptance of plate tectonics was a testament to the **integration of diverse geophysical data**, where seismic, magnetic, and gravity data were synthesized (e.g., McKenzie and Parker, 1967) to create a cohesive and powerful model of Earth's dynamic, moving lithospheric plates.

Environmental and Engineering Geophysics: As the century progressed, geophysical methods found new and critical applications beyond traditional resource exploration. The nascent fields of **environmental and engineering geophysics** emerged, utilizing established techniques for addressing societal challenges. Early applications included investigations into **groundwater contamination**, characterizing hazardous waste sites like **landfills**, and conducting **engineering site investigations** for infrastructure projects such as foundation studies, demonstrating the growing versatility and societal relevance of geophysics.

2.3 Contemporary Geophysics

Contemporary geophysics stands at the forefront of scientific inquiry, leveraging cutting-edge technologies, immense computational power, and multidisciplinary approaches to unravel Earth's complexities and address pressing global challenges. This modern era is characterized by an unprecedented ability to acquire, process, and interpret vast quantities of diverse geophysical data.

Advanced Data Acquisition and Processing: The bedrock of modern geophysics lies in its sophisticated data collection methods. **High-resolution seismic** techniques, including 3D and the time-lapse 4D surveys, have revolutionized our understanding of the subsurface, particularly for energy exploration and reservoir monitoring. The development of **full-waveform inversion (FWI)** has enabled unprecedented detail in subsurface imaging by utilizing the entire seismic waveform. This wealth of data necessitates equally advanced processing algorithms to extract meaningful information. Beyond active source methods, **satellite geodesy and remote sensing** offer a global perspective. Technologies like **GPS** provide precise positional data, while **InSAR (Interferometric Synthetic Aperture Radar)** allows for the detection and monitoring of ground deformation due to phenomena such as subsidence, landslides, and volcanic unrest. Dedicated satellite gravity missions, exemplified by **GRACE** and **GOCE**, have provided invaluable insights into mass transport, including changes in ice sheets, ocean currents, and crucial water resources (e.g., Tapley et al., 2004). More recently, **Unmanned Aerial Vehicles (UAVs/Drones)** are increasingly deployed for targeted magnetic, radiometric, and even seismic surveys, offering flexibility and access to difficult or hazardous terrains.

Computational Geophysics and Data Science: The sheer volume and complexity of geophysical data have propelled computational geophysics and data science to central roles. **Numerical modeling and inversion** allow scientists to simulate complex geophysical processes, from earthquake rupture dynamics to fluid flow in porous media. Advanced inversion techniques enable the extraction of more detailed and accurate subsurface information from observed data. The transformative power of **Machine Learning (ML) and Artificial Intelligence (AI)** is now being harnessed across various geophysical domains. AI/ML algorithms are employed for automated seismic interpretation, anomaly detection in vast datasets, optimizing data processing workflows, and even for predictive modeling, such as in earthquake forecasting or guiding resource exploration efforts. The handling and interpretation of these enormous datasets necessitate robust **Big Data Analytics** frameworks.

Multidisciplinary Applications and Societal Impact: The applications of contemporary geophysics are incredibly diverse, with profound societal impact. In **climate change research**, geophysical methods are crucial for monitoring ice sheet dynamics, sea-level rise, permafrost thaw, and groundwater depletion, providing critical data for climate models and mitigation strategies. Geophysics plays a vital role in **natural hazard mitigation**, informing advanced seismic hazard assessments, real-time volcano monitoring, and landslide prediction through the integration of various geophysical data streams. The global shift towards sustainable energy has seen a renewed focus on geophysics for **geothermal energy exploration**, where techniques are used to locate and characterize viable geothermal reservoirs. Furthermore, in response to growing global demands, geophysics is central to the sustainable exploration of **critical minerals and groundwater resources**. **Near-surface geophysics** continues to refine its methods for diverse environmental and engineering challenges, including the assessment of brownfield sites, detection of unexploded ordnance (UXO), monitoring of critical infrastructure, and even precision agriculture applications. Beyond Earth, **planetary geophysics** applies terrestrial principles and techniques to study other celestial bodies, as exemplified by the InSight mission's seismic investigations on Mars and ongoing lunar gravity mapping efforts.

Future Trends and Challenges: The future of geophysics is characterized by several key trends. The **integration of "smart" technologies**, including IoT (Internet of Things) sensors and real-time data streaming, will enable continuous monitoring of Earth processes. There is a growing emphasis on **open science and data sharing**, fostering collaborative research and making vast datasets more accessible to the global scientific community. Ultimately, geophysics will continue to play a critical role in addressing global grand challenges, including facilitating the **energy transition**, ensuring **water security**, and enhancing societal **resilience to natural hazards**.

2.4 Traditional Methods of Geophysics

The 20th century witnessed the maturation of several fundamental geophysical methods that continue to form the bedrock of applied geophysics. While continually refined with advancements in technology and computational power, their underlying principles remain indispensable for understanding the Earth's subsurface.

Seismic Methods

Seismic methods involve generating elastic waves and analyzing their propagation through the Earth. They are arguably the most widely used geophysical techniques, particularly in hydrocarbon exploration.

- **Seismic Reflection:** This method operates on the principle of generating seismic waves (e.g., from explosives or vibroseis trucks) and recording the reflections of these waves from subsurface geological interfaces using geophones. By analyzing the travel times and amplitudes of these reflections, detailed images of subsurface structures can be created (Dobrin, 1960). Its early development was driven by the burgeoning oil industry in the early 20th century, with pioneers like Ludger Mintrop making significant contributions. Today, it remains the primary method for hydrocarbon exploration, providing high-resolution structural maps of faults and folds, and enabling detailed stratigraphic interpretation. However, it requires significant energy sources, is sensitive to environmental noise, and can be challenging in geologically complex areas.
- **Seismic Refraction:** Unlike reflection, seismic refraction measures the travel times of critically refracted seismic waves that travel along subsurface layers. This allows for the determination of layer velocities and depths (Telford et al., 1990). While often predating reflection seismology in some applications, it is particularly suited for shallower investigations. Its applications include bedrock mapping for construction, "rippability" studies to assess excavation difficulty, shallow groundwater exploration, and engineering site characterization. Limitations include an assumption of increasing velocity with depth, limited depth penetration compared to reflection, and difficulties in accurately characterizing dipping layers or velocity inversions.

Gravity Method

The gravity method measures subtle variations in the Earth's gravitational field that are caused by differences in the density of subsurface rocks (Parasnis, 1997). The instrumentation has evolved significantly, from early pendulums and torsion balances to the highly precise spring-based gravimeters used today. Data acquisition is typically conducted through land-based surveys, taking measurements at discrete stations, though early airborne surveys were also utilized for regional mapping. Applications are diverse, ranging from regional geological mapping to petroleum exploration (identifying low-density salt domes) and mineral exploration (locating high-density massive sulfide deposits). It also plays a crucial role in large-scale crustal studies. A significant limitation is the non-uniqueness of interpretation, meaning multiple density distributions can produce the same observed gravity anomaly. It is also highly sensitive to topography and requires extensive corrections.

Magnetic Method

The magnetic method involves measuring variations in the Earth's magnetic field that are caused by differences in the magnetic susceptibility of subsurface rocks and minerals (Blakely, 1995). Early instruments like dip needles and compasses paved the way for more sophisticated devices such as proton precession and fluxgate magnetometers. Data acquisition is carried out through ground surveys and, notably, extensively through airborne platforms (aeromagnetic surveys), which allow for rapid and cost-effective regional coverage. Key applications include mineral exploration (e.g., iron ore, kimberlites), geological mapping (identifying igneous intrusions and faults), unexploded ordnance (UXO) detection, and even archaeological investigations. However, the method can be affected by "cultural noise" from man-made structures like power lines and fences, and interpretation can be ambiguous regarding the exact depth and geometry of the magnetic source. Complex magnetization models can also pose challenges.

Electrical Methods (DC Resistivity & IP)

Electrical methods involve inducing or measuring electrical currents in the ground to infer subsurface properties.

- **Direct Current (DC) Resistivity:** This method involves injecting direct current into the ground through a pair of electrodes and measuring the resulting potential difference between another pair of electrodes. This allows for the determination of the subsurface's electrical resistivity (Loke, 2000), which varies depending on rock type, fluid content, and porosity. Early development focused on simple electrode configurations like Wenner and Schlumberger arrays for both vertical soundings and horizontal profiling. Applications are widespread, including groundwater exploration, environmental investigations (e.g., mapping contaminant plumes), geotechnical site characterization, and mineral exploration. Limitations include sensitivity to cultural noise, limited depth penetration, and susceptibility to poor electrode contact.
- **Induced Polarization (IP):** IP measures the transient voltage decay after the cessation of current injection, indicating the "chargeability" of the subsurface (Sumner, 1976). This phenomenon occurs due to electrochemical reactions at mineral-water interfaces. Primarily used for disseminated sulfide mineral exploration, it also finds applications in environmental studies, such as mapping clay-rich zones or certain types of contamination. Interpretation can be complex, and the method shares some limitations with DC resistivity.

Electromagnetic (EM) Methods (Early Applications)

Electromagnetic (EM) methods work by inducing eddy currents in the subsurface using an electromagnetic field and then measuring the secondary magnetic field generated by these currents (McNeill, 1980). Early instruments included systems like Slingram, VLF (Very Low Frequency), and initial time-domain EM (TEM) systems. Applications span mineral exploration (particularly for conductive ore bodies), groundwater mapping, and environmental studies. The depth of penetration for EM methods is dependent on the frequency used (for frequency-domain systems) or the time window analyzed (for time-domain systems). They can be sensitive to highly conductive overburden and present challenges in areas with significant lateral variations in subsurface conductivity.

2.5 Modern Methods of Geophysics

The dawn of the 21st century has seen a dramatic transformation in geophysical methods, largely propelled by exponential increases in computational power, the advent of sophisticated satellite technology, and innovative sensor development. These advancements have led to unprecedented resolution, deeper penetration capabilities, and a burgeoning array of new applications across diverse fields.

Advanced Seismic Methods

Seismic methods, already foundational, have undergone revolutionary advancements:

- **3D and 4D Seismic:** This represents a leap from traditional 2D profiles to the acquisition of dense 3D grids of seismic data, providing a volumetric understanding of the subsurface. **4D seismic** (time-lapse) takes this a step further by repeatedly acquiring 3D data over time to monitor dynamic changes, such as fluid movement within hydrocarbon reservoirs (Lumley, 2010). This has revolutionized hydrocarbon exploration and reservoir management, enabling detailed reservoir characterization, enhanced oil recovery (EOR) monitoring, CO2 sequestration monitoring, and the imaging of highly complex geological structures.
- **Full-Waveform Inversion (FWI):** FWI is a computationally intensive, iterative method that uses the *entire* seismic waveform – both amplitudes and phases – to construct highly detailed and accurate subsurface velocity models (Virieux and Operto, 2009). Its impact lies in significantly improving imaging resolution and accuracy, making it invaluable for

sub-salt imaging, delineating complex geological structures, and facilitating quantitative reservoir characterization.

- **Seismic Attributes and Machine Learning:** Modern seismic interpretation heavily relies on the derivation of various attributes (e.g., coherence, curvature, spectral decomposition) from seismic data to highlight specific geological features. Crucially, the integration of **machine learning** algorithms with these attributes allows for automated interpretation, anomaly detection, and the identification of subtle features that might be missed by human interpreters. This speeds up interpretation workflows, reduces human bias, and enhances the overall understanding of subsurface geology, aiding in facies analysis, fault detection, stratigraphic interpretation, and prospect generation.

Satellite-Based Geophysics and Remote Sensing

Space-borne technologies have opened up entirely new avenues for geophysical observation:

- **GPS/GNSS (Global Navigation Satellite Systems):** By leveraging signals from satellite constellations, GPS and other GNSS provide high-precision measurements of ground deformation. This allows for continuous monitoring of tectonic plate movements, volcanic inflation and deflation, and land subsidence (Blewitt et al., 2018). Its applications are critical for geodynamics, natural hazard monitoring, and assessing infrastructure deformation.
- **InSAR (Interferometric Synthetic Aperture Radar):** InSAR uses radar signals emitted from satellites to detect subtle changes in ground elevation over vast areas (Massonnet and Feigl, 1998). This technique has become indispensable for monitoring landslides, urban subsidence, post-seismic deformation, glacier dynamics, and volcanic activity, providing millimeter-scale precision.
- **Satellite Gravity (GRACE, GOCE):** Missions like GRACE (Gravity Recovery and Climate Experiment) and GOCE (Gravity field and steady-state Ocean Circulation Explorer) have revolutionized our ability to measure Earth's time-varying gravity field with unprecedented accuracy. These data are vital for monitoring large-scale hydrological processes like changes in groundwater storage, assessing ice sheet mass balance, and understanding ocean currents (Tapley et al., 2004).

Advanced Electromagnetic (EM) Methods

EM methods have seen significant enhancements in resolution, depth penetration, and coverage:

- **Airborne EM (AEM) Systems:** Modern AEM systems, deployed from fixed-wing aircraft or helicopters, allow for rapid, large-area coverage with enhanced depth penetration. These systems often incorporate multi-frequency and multi-component measurements, providing more comprehensive data. Applications include regional geological mapping, mineral exploration, groundwater investigations, and environmental surveys.
- **Controlled-Source Electromagnetic (CSEM) and Magnetotellurics (MT):** CSEM employs an artificial source to actively probe deep electrical resistivity structures in the Earth, particularly useful in marine environments. MT, in contrast, utilizes naturally occurring telluric currents and variations in Earth's magnetic field to image deeper resistivity structures. Both methods are crucial for hydrocarbon exploration (e.g., beneath challenging salt bodies), geothermal exploration, and deep crustal imaging, including for unconventional resources and deep ore deposits.

Ground Penetrating Radar (GPR)

GPR emits high-frequency electromagnetic pulses into the ground and records reflections from subsurface interfaces (Daniels, 2004). Its primary impact is providing high-resolution imaging of shallow subsurface features. Applications are diverse and include utility mapping (pipes, cables), archaeological investigations, concrete imaging, unexploded ordnance (UXO) detection, and various near-surface environmental studies (e.g., soil moisture content, contaminant plumes). A key limitation is its limited depth penetration in highly conductive materials like clays and saline water, as well as signal attenuation.

Borehole Geophysics (Well Logging)

Borehole geophysics, or well logging, involves lowering various geophysical sensors into boreholes to measure the physical properties of formations *in situ* (e.g., resistivity, natural gamma ray, density, sonic velocity). This provides crucial "ground truth" and high-resolution vertical profiles of the subsurface. It is an indispensable tool for formation evaluation in the oil and gas industry, hydrogeology, geotechnical studies, and mineral exploration, directly correlating geophysical measurements with geological samples.

Distributed Acoustic Sensing (DAS)

DAS is a relatively new and transformative technology that utilizes fiber optic cables as continuous arrays of seismic sensors. By sending laser pulses down the fiber and analyzing the backscattered light, DAS can detect minute strains caused by acoustic waves, effectively turning the entire cable into a dense seismic array. This enables continuous and high-density acoustic monitoring, transforming seismic data acquisition and allowing for real-time monitoring of various phenomena. Applications are rapidly expanding and include pipeline and railway monitoring, microseismic monitoring (e.g., during hydraulic fracturing operations), and enhanced earthquake detection.

CHAPTER 3: THE COMPONENTS OF APPLIED GEOPHYSICS

This chapter focusses on the practical aspects of geophysics, exploring how scientific principles and technological innovations have been harnessed to solve real-world problems. We will trace the evolution of applied geophysics from its nascent stages to its sophisticated contemporary forms, highlighting the methods and technologies that define its various eras.

3.1 Early History of Applied Geophysics

The journey of applied geophysics, the practical use of geophysical principles to understand the Earth, began long before it was formally recognized as a scientific discipline. Its early history is characterized by a fascinating blend of intuitive observations, rudimentary techniques, and eventually, the dawn of systematic, instrument-based investigations driven by the imperative of resource exploration.

Pre-Scientific Applications and Intuition

Even in ancient times, humans implicitly engaged in "geophysical" observations. Ancient miners, for instance, were acutely aware of changes in rock types, water flow, and other surface indicators that hinted at valuable mineral deposits. While not scientific, practices like dowsing rods – an early, albeit unscientific, attempt to "locate" resources – reflect humanity's innate desire to find what lies hidden beneath the surface. Similarly, the early use of the magnetic compass by the Chinese for navigation, while not directly applied geophysics in the modern sense, represented the earliest practical application of Earth's magnetic field. This set a crucial precedent for harnessing natural geophysical phenomena for tangible, practical ends.

The Dawn of Systematic Application (Late 19th - Early 20th Century)

The late 19th and early 20th centuries marked a pivotal shift from intuition to systematic, instrument-based geophysical exploration, primarily spurred by the growing demand for natural resources.

- **Magnetic Prospecting:** This was one of the very first truly "applied" geophysical methods. Early instruments, such as dip needles and simple magnetometers, were employed to detect variations in the Earth's magnetic field caused by magnetic ore bodies like iron ore and magnetite (Nettleton, 1940). This method quickly found success, with early successful applications in Sweden and the United States demonstrating its efficacy in delineating significant iron deposits.

- **Gravimetric Prospecting:** The introduction of the Eötvös torsion balance revolutionized gravimetric prospecting. This sophisticated instrument allowed for the quantitative measurement of subtle density variations in the subsurface (LaCoste, 1941). Its immediate impact was seen in the detection of salt domes in the Gulf Coast region of the United States, which were recognized as important traps for oil and gas. This marked the beginning of quantitative gravimetric surveys in the field, moving beyond mere qualitative observations.
- **Early Electrical Methods:** Primitive **electrical resistivity measurements** also emerged in this period, primarily aimed at locating ore bodies or groundwater. The work of the **Schlumberger brothers in the 1910s and 1920s** was particularly influential, even if the early applications were more empirical than theoretically fully developed (Allaud and Allaud, 1977). These initial applications were largely concentrated in mining for conductive sulfide deposits and, to a lesser extent, for groundwater exploration.
- **Seismic Sounding (Pre-Reflection):** The initial experiments with seismic waves for subsurface mapping primarily utilized refraction principles to determine the depth to bedrock or to identify large structural features like salt domes. Ludger Mintrop's pioneering work in Germany and later in the U.S. around World War I was instrumental in demonstrating the potential of seismic methods for oil exploration (Sheriff and Geldart, 1995). At this early stage, the scope of seismic sounding was limited to structural features and did not yet provide the detailed stratigraphic imaging that would characterize later seismic reflection techniques.

3.2 The Rise of Modern Applied Geophysics

The mid-20th century marked a pivotal period for applied geophysics, transforming it from a collection of observational techniques into a sophisticated, industrialized discipline. This remarkable evolution was largely propelled by the burgeoning demands of the petroleum industry and significant innovations stemming from wartime research.

The Age of Oil and Gas Exploration: Seismic Reflection Dominance

The quest for hydrocarbons became the primary driver for geophysical advancements. Seismic Reflection emerged as the undisputed king of exploration tools. Its dominance was solidified by crucial methodological and technological refinements. The development of multi-channel recording allowed for the simultaneous acquisition of data from multiple geophones, dramatically improving data quality. The introduction of Common Depth Point (CDP) stacking revolutionized noise reduction and enhanced signal-to-noise ratios, leading to clearer subsurface images. Crucially, the shift to digital processing (Robinson and Treitel, 1980) unlocked the potential for complex algorithms for noise attenuation, deconvolution, and migration, moving seismic interpretation from art to a more rigorous science. These advancements were underpinned by post-WWII technological drivers, including rapid progress in electronics, the nascent field of computing, and sophisticated instrumentation. This period also saw a global expansion of geophysics laboratories and service companies, proliferating worldwide to meet the insatiable energy demands.

Evolution of Gravimetry and Magnetics

Gravity and magnetic methods also underwent significant advancements, particularly in their acquisition efficiency. The development of airborne magnetometers and, later, airborne gravimeters revolutionized reconnaissance surveys, allowing for rapid and cost-effective regional coverage (e.g., Jensen, 1961). This significantly reduced the time and cost of initial exploration. Instrumentation became more refined with the advent of highly stable and precise LaCoste & Romberg gravimeters and the ubiquitous proton precession magnetometers. This improved data quality paved the way for more quantitative interpretation, with the development of rigorous forward modeling and inversion techniques for gravity and magnetic data, moving beyond qualitative anomaly recognition.

Advancements in Electrical and Electromagnetic Methods

Electrical and electromagnetic techniques also saw substantial progress, expanding their utility beyond basic resistivity surveys. The systematic development and application of Induced Polarization (IP) became particularly significant for disseminated sulfide mineral exploration (Sumner, 1976). IP provided a powerful tool for finding ore deposits that lacked strong magnetic or gravity signatures, opening up new frontiers in mining. Furthermore, the introduction of Time-Domain EM (TEM) systems offered advantages in terms of deeper penetration and better resolution in conductive environments compared to some of the earlier frequency-domain methods, further enhancing their capabilities for mineral and groundwater exploration.

The Role of Computing

Perhaps the single most transformative factor was the pervasive role of computing. The shift from analog to digital data acquisition and processing was a game-changer. This allowed for the implementation of complex algorithms for noise reduction, signal enhancement, deconvolution (to remove the effects of the Earth's filter), and migration (to correctly position reflections in their true subsurface locations) in seismic processing. Beyond seismic, the development of basic numerical models for interpreting geophysical anomalies in gravity, magnetics, and electrical methods laid the groundwork for the sophisticated modeling and inversion capabilities we see today.

Diversification Beyond Oil and Gas

While petroleum exploration remained a dominant force, the advancements in geophysical methods led to their diversification beyond oil and gas. The mining industry continued and expanded its use of seismic, gravity, magnetic, and various electrical methods for the efficient exploration and delineation of diverse ore deposits. The increasing demand for fresh water also spurred the increased application of resistivity and early EM methods for mapping aquifers and understanding groundwater flow. Furthermore, a nascent use of geophysics emerged for environmental and engineering applications, primarily for shallow site characterization, identifying buried utilities, and addressing early environmental concerns. This diversification marked the beginning of geophysics' broader impact on society.

3.3 Contemporary Applied Geophysics

Contemporary applied geophysics stands at the forefront of scientific and technological innovation, characterized by an unprecedented level of data integration, advanced computational methods, multidisciplinary collaboration, and a broadening focus beyond traditional resource exploration to address pressing global challenges.

Data Integration and Multidisciplinary Approaches

The modern geophysicist rarely works in isolation. Synergistic interpretation is now common practice, integrating data from diverse geophysical methods (seismic, gravity, magnetic, electromagnetic, well logs) with geological, geochemical, and remote sensing data. This holistic approach provides a far more comprehensive understanding of the subsurface than any single method could achieve. A transformative force in this integration is the increasing use of Machine Learning (ML) and Artificial Intelligence (AI) algorithms. These technologies are revolutionizing data processing, interpretation, and anomaly detection. They enable automated facies classification, predictive modeling for reservoir properties or mineral prospectivity, and unlock insights from massive datasets that were previously intractable (Chen et al., 2020).

High-Resolution and Time-Lapse Imaging

The demand for ever more detailed and dynamic views of the subsurface has driven significant advancements. 4D seismic, or time-lapse seismic, is now routinely applied to monitor fluid movement in hydrocarbon reservoirs, track the injection and containment of CO2 in sequestration sites, and manage geothermal fields (Lumley, 2010). This allows for dynamic reservoir management and environmental monitoring. Concurrently, high-resolution near-surface geophysics has seen substantial development, particularly with advanced Ground Penetrating Radar (GPR), shallow seismic, and sophisticated EM systems. These tools provide precise imaging of the top tens of meters, critical for environmental investigations, geotechnical site characterization, and crucial for urban infrastructure mapping and archaeological applications.

Big Data and Computational Power

Modern geophysical surveys generate massive datasets, often reaching petabytes from large-scale acquisitions such as ocean bottom seismic and wide-azimuth surveys. Managing and processing such volumes is only possible through reliance on High-Performance Computing (HPC), including supercomputers and cloud computing resources. These computational capabilities enable highly intensive tasks like Full-Waveform Inversion (FWI), which uses the entire seismic waveform to build extremely accurate subsurface velocity models, and complex numerical simulations that enhance understanding of wave propagation and subsurface responses.

Emerging Sensor Technologies and Platforms

Innovation in sensor technology and deployment platforms is continually expanding the capabilities of geophysics. Distributed Acoustic Sensing (DAS), which utilizes fiber optic cables as continuous seismic arrays, is emerging as a game-changer for pipeline monitoring, unconventional reservoir characterization, and urban seismology (Parker et al., 2014). The widespread adoption of Unmanned Aerial Vehicles (UAVs/Drones) has revolutionized data acquisition, enabling rapid and cost-effective magnetic, radiometric, lidar, and even mini-seismic surveys in challenging or remote terrains. Furthermore, Satellite Geodesy and Remote Sensing are now routine tools, with GPS/GNSS and InSAR (Interferometric Synthetic Aperture Radar) providing highly precise measurements of subtle ground deformation related to natural hazards (earthquakes, volcanoes, landslides), resource extraction-induced subsidence, and cryosphere dynamics, such as the movement of glaciers and ice sheets (Massonnet and Feigl, 1998; Tapley et al., 2004).

Addressing Global Challenges

Contemporary applied geophysics is increasingly pivotal in addressing critical global challenges beyond its traditional focus on fossil fuels:

- **Energy Transition:** Geophysics plays a crucial role in exploring for and developing geothermal energy resources, identifying suitable sites for subsurface storage of hydrogen and CO2, and optimizing the recovery from unconventional resources.
- **Water Security:** High-resolution mapping of aquifers, monitoring groundwater depletion, and detecting saltwater intrusion are essential applications for ensuring sustainable water supplies.
- **Natural Hazard Mitigation:** Advanced geophysical techniques contribute significantly to seismic hazard assessment, real-time volcano monitoring, comprehensive landslide risk assessment, and the development of early warning systems for various natural disasters.
- **Critical Minerals:** With the surging demand for technologies related to renewable energy and electric vehicles, geophysics is playing a crucial role in exploring for new deposits of battery minerals and rare earth elements.

- **Environmental Monitoring:** Geophysics provides invaluable tools for the characterization of contaminated sites, waste disposal monitoring, and understanding the impacts of climate change, such as permafrost degradation studies.

This broad scope underscores how contemporary applied geophysics is not just about finding resources but about understanding, monitoring, and managing our planet for a sustainable future.

3.4 Traditional Methods Used in Applied Geophysics

The suite of geophysical methods often termed "traditional" represents the fundamental techniques that remain the workhorses of applied geophysics. While continuously refined by technological advancements and integrated with modern computational power, their underlying principles and diverse applications across various sectors underscore their enduring importance.

Seismic Reflection

Seismic reflection is a cornerstone of applied geophysics, providing high-resolution images of the subsurface. Its fundamentals involve generating seismic waves (acoustic energy) and recording their reflections from geological interfaces within the Earth. Key concepts include understanding seismic wave propagation (compressional and shear waves), the principles of reflection and refraction at impedance contrasts, and the careful planning of acquisition geometry using source-receiver layouts. Common Mid-Point (CMP) gathering is a crucial technique that allows for the stacking of multiple traces to enhance signal and attenuate noise. Basic **processing steps** like Normal Moveout (NMO) correction, stacking, and migration (Yilmaz, 2001) are essential to transform raw data into interpretable subsurface images.

Applications of seismic reflection are broad and impactful. It is the dominant method for hydrocarbon exploration, providing detailed structural mapping of traps, stratigraphic analysis for reservoir identification, and increasingly, reservoir characterization to understand fluid distribution and rock properties. Beyond oil and gas, it's vital for CO2 storage assessment and geothermal exploration. Variations include 2D surveys (profiles), 3D surveys (volumetric imaging), and early forms of 4D surveys (time-lapse monitoring of changes over time).

Gravity Method

The gravity method relies on measuring minute variations in the Earth's gravitational field, which are caused by differences in subsurface rock density. Its fundamentals involve understanding the inverse square law of gravity, how density contrasts create gravitational anomalies, and the necessity of applying Bouguer and free-air corrections (Telford et al., 1990) to isolate anomalies caused by subsurface geology from those due to elevation and topography.

Instrumentation primarily involves highly sensitive gravimeters, deployed in land-based, marine (shipborne), and airborne platforms. Applications include regional geological mapping to delineate large-scale features like sedimentary basins and structural uplifts, salt dome exploration (due to their low density), and the detection of massive sulfide deposits (due to their high density).

Magnetic Method

The magnetic method measures variations in the Earth's total magnetic field caused by differences in the magnetic susceptibility of subsurface rocks and minerals. Fundamentals include understanding the Earth's main magnetic field, how magnetic anomalies are generated by contrasts in magnetic susceptibility, and distinguishing between induced magnetization (due to the Earth's current field) and remanent magnetization (fossil magnetization from past geological eras) (Blakely, 1995).

Instrumentation has evolved from simple compasses to highly sensitive proton precession and cesium vapor magnetometers. Data acquisition is conducted from various platforms, including ground surveys, extensive airborne (aeromagnetic) surveys for rapid coverage, and marine surveys. Applications are diverse, encompassing mineral exploration (e.g., iron ore, kimberlites), geological mapping (identifying igneous intrusions, faults, and dykes), unexploded ordnance (UXO) detection, and archaeological investigations for mapping buried structures.

Electrical Resistivity Method (DC and AC Resistivity)

The electrical resistivity method involves injecting electrical current into the ground and measuring the resulting potential differences to infer the subsurface's electrical resistivity. Fundamentals revolve around Ohm's Law and the concept of apparent resistivity, which is then used to deduce true resistivity. Various common array configurations (e.g., Wenner, Schlumberger, dipole-dipole) (Loke, 2000) are employed to achieve different depths of investigation and resolutions. Both Direct Current (DC) and Alternating Current (AC) resistivity techniques are utilized.

Applications are widespread, including groundwater exploration (mapping aquifers), environmental investigations (e.g., delineating contaminant plumes, mapping old landfills), geotechnical site characterization (determining bedrock depth, identifying soil types), and archaeological mapping for locating buried features.

Induced Polarization (IP) Method

The Induced Polarization (IP) method measures the subsurface's ability to store an electrical charge, known as chargeability. This phenomenon is primarily due to electrochemical phenomena occurring at mineral-fluid interfaces (Sumner, 1976), particularly in the presence of disseminated metallic sulfides.

Its primary application is in disseminated sulfide mineral exploration (e.g., porphyry copper deposits), where it can detect small, finely dispersed mineral grains that might not be visible to other methods. It also finds use in mapping clay-rich zones and certain types of environmental contamination.

Electromagnetic (EM) Methods (Ground-based and Airborne)

Electromagnetic (EM) methods involve inducing eddy currents in the subsurface using a primary electromagnetic field and then measuring the secondary magnetic field generated by these induced currents. Fundamentals include principles of electromagnetic induction, the concept of primary and secondary fields, and the relationship between subsurface conductivity and the measured response. Various EM systems exist, operating in both the frequency-domain (varying transmitted frequency) and time-domain (measuring decay of secondary field after a pulse), and very low frequency (VLF) methods (McNeill, 1980).

Applications are significant in mineral exploration (especially for conductive ore bodies like massive sulfides), groundwater exploration (mapping water-bearing zones), saline intrusion mapping in coastal areas, and a range of environmental site characterization tasks.

3.5 Modern Methods of Applied Geophysics

Recent decades have witnessed a paradigm shift in applied geophysics, driven by exponential advancements in computing power, sophisticated sensor technology, and a growing emphasis on interdisciplinary integration. These innovations have propelled the field into an era of unprecedented subsurface imaging resolution and continuous Earth monitoring.

Advanced Seismic Imaging and Characterization

Seismic methods, already foundational, have undergone revolutionary advancements:

- **Full-Waveform Inversion (FWI):** FWI represents a leap forward in seismic imaging. Instead of just using travel times, it leverages the *entire* seismic waveform (amplitude and phase) to iteratively construct highly accurate and detailed velocity models of the subsurface (Virieux and Operto, 2009). This provides vastly superior resolution compared to conventional migration techniques. FWI is now crucial for high-resolution imaging in complex geological settings (e.g., beneath salt bodies or basalt layers), detailed reservoir characterization, and quantitative interpretation for predicting fluid flow and rock properties.
- **Distributed Acoustic Sensing (DAS):** DAS is a groundbreaking technology that transforms standard fiber optic cables into dense, continuous arrays of acoustic sensors by measuring Rayleigh backscatter (Parker et al., 2014). This allows for continuous, spatially extensive seismic measurements along the entire length of the fiber. Applications are diverse, including pipeline integrity monitoring, vertical seismic profiling (VSP) in wellbores, microseismic monitoring (critical for hydraulic fracturing and geothermal energy), and passive seismic monitoring for earthquake detection and urban noise characterization.
- **Seismic Attributes and Machine Learning/Deep Learning:** The extraction of seismic attributes—quantitative measures from seismic data such as coherence, curvature, spectral decomposition, impedance, and porosity—has become standard practice to highlight subtle geological features. The integration of Artificial Intelligence (AI), Machine Learning (ML), and Deep Learning (DL) has revolutionized interpretation. These powerful algorithms enable automated fault detection, facies classification, accurate reservoir property prediction, and accelerated seismic interpretation. This impact is profound,

significantly speeding up interpretation workflows, enhancing accuracy, and uncovering subtle patterns within massive seismic datasets that might be missed by human interpreters.

Satellite-Based Geophysics

The advent of satellite technology has opened up an entirely new dimension for geophysical observation, providing unprecedented spatial coverage and temporal resolution for monitoring Earth's dynamic processes:

- **InSAR (Interferometric Synthetic Aperture Radar):** InSAR uses the phase differences between two radar images acquired from space at different times to measure millimeter-scale ground deformation (Massonnet and Feigl, 1998). This powerful technique is widely applied for monitoring landslides, volcanic inflation/deflation cycles, land subsidence due to groundwater withdrawal or hydrocarbon production, earthquake-induced deformation, and assessing the stability of critical infrastructure.
- **Satellite Gravimetry (GRACE/GRACE-FO, GOCE):** Missions like GRACE (Gravity Recovery and Climate Experiment), GRACE-FO (Follow-On), and GOCE (Gravity field and steady-state Ocean Circulation Explorer) measure subtle changes in Earth's gravity field over time. These changes reflect the redistribution of mass within the Earth system (Tapley et al., 2004). This allows for monitoring continental water storage changes (groundwater, snow, ice), ice sheet mass balance, and ocean bottom pressure changes, providing critical data for climate science and hydrology.
- **GNSS (Global Navigation Satellite Systems) / GPS:** Beyond basic navigation, highly precise positioning measurements from GNSS (including GPS, GLONASS, Galileo, BeiDou) are used to track crustal deformation, tectonic plate motions, and localized ground movement with exquisite accuracy (Blewitt et al., 2018). This capability is indispensable for earthquake studies (measuring co-seismic and post-seismic deformation), monitoring volcano deformation, tracking landslide movements, and observing post-seismic rebound.

Ground Penetrating Radar (GPR)

GPR has matured into a high-resolution shallow imaging tool. It operates by emitting high-frequency electromagnetic waves into the subsurface and recording reflections from contrasts in dielectric properties (Daniels, 2004). GPR provides unparalleled detail for shallow targets, making it invaluable for utility mapping (identifying buried pipes and cables), archaeological surveys (locating buried structures and artifacts), concrete inspection, unexploded ordnance (UXO) detection, burial site mapping, and environmental site characterization (e.g., mapping soil moisture or buried drums).

Borehole Geophysics / Well Logging (Advanced Techniques)

While well logging has been around for decades, modern advancements have significantly enhanced its capabilities. These involve the collection of highly detailed physical property measurements directly within boreholes. Advanced techniques include Nuclear Magnetic Resonance (NMR) logging for direct fluid typing and porosity estimation, acoustic imaging for detailed structural and fracture analysis of the borehole wall, and micro-resistivity imaging for high-resolution imaging of geological features and fractures. These advanced logs are critical for detailed formation evaluation, precise petrophysical property estimation, accurate fracture detection, and monitoring fluid saturation changes in hydrocarbon and geothermal reservoirs.

Unmanned Aerial Vehicle (UAV) Geophysics

The integration of miniature, lightweight geophysical sensors onto Unmanned Aerial Vehicles (UAVs or drones) represents a rapidly expanding frontier. Drones can carry magnetometers, gamma-ray spectrometers, and lightweight GPR/EM systems, enabling rapid, cost-effective, and high-resolution surveys in remote or hard-to-access areas. Applications include targeted mineral exploration, environmental mapping (e.g., detecting groundwater contamination), efficient UXO detection, infrastructure inspection (e.g., dams, bridges), and precision agriculture (e.g., mapping soil properties).

CHAPTER 4: APPLIED GEOPHYSICS IN MINING

Applied geophysics plays a critical role throughout the entire lifecycle of a mining project, from grassroots exploration to resource definition, mine planning, operational monitoring, and post-closure environmental management. This chapter will comprehensively explore the application of various geophysical methods in the mining sector, emphasizing their unique strengths and limitations in different geological settings and for various deposit types.

4.1 Introduction to Mining Geophysics

The quest for mineral resources has driven human innovation for millennia. From the earliest rudimentary prospecting based on surface outcrops and stream sediments to the sophisticated methodologies of today, the challenge has always been to efficiently and effectively locate valuable deposits hidden beneath the Earth's surface. This is where applied geophysics becomes indispensable.

Applied geophysics in mining is the specialized field that utilizes measurements of Earth's physical properties to infer the presence, location, and characteristics of mineral deposits and the geological structures that control them. It goes beyond simple observation, employing precise instrumentation and advanced analytical techniques to "see" into the subsurface without direct excavation, offering a non-invasive and often more cost-effective alternative or complement to traditional drilling.

The integration of geophysical methods is not a standalone activity but a crucial component throughout the entire mining lifecycle, influencing decisions from initial grassroots exploration to final mine closure and environmental stewardship. Its strategic application helps to significantly reduce exploration risk, optimize resource definition, enhance operational efficiency, and ensure environmental responsibility.

4.1.1 Role of Geophysics in the Mining Lifecycle

Understanding the specific contributions of geophysics at each stage of a mining project is key to appreciating its comprehensive value:

- **Reconnaissance (Greenfield Exploration):** This is the earliest stage, typically involving large, often underexplored areas. Here, geophysics acts as a powerful regional screening tool. Cost-effective airborne surveys (e.g., magnetics, radiometrics, airborne EM) are frequently employed to:

- Identify large-scale geological features (e.g., crustal structures, major lithological contacts, buried basins) that are favorable for mineralization.
- Pinpoint areas of anomalous physical properties that could indicate the presence of deeply buried or subtle mineral systems, even under extensive cover (e.g., sediments, glacial till).
- Prioritize smaller, more prospective zones for follow-up, thereby minimizing the vast expense of widespread ground-based work or drilling. This phase aims to narrow down millions of square kilometers to a few prospective hundreds or thousands.

- **Target Generation (Brownfield Exploration):** Once a promising region is identified, or within existing mineral belts and near operating mines, the focus shifts to delineating specific drill targets. This stage demands higher resolution and often involves ground-based geophysical surveys (e.g., detailed magnetics, IP, resistivity, ground EM, gravity). Geophysics helps to:
  - Map the precise geometry of known or inferred ore bodies at depth.
  - Delineate associated alteration zones or structural controls (faults, shear zones) that act as conduits for mineralization.
  - Identify "blind" deposits – those with no surface expression – by detecting their characteristic geophysical signatures.
  - Optimize drill hole placement, ensuring that expensive drilling is directed to the most promising locations, maximizing the probability of discovery and minimizing unnecessary expenditure.
- **Resource Definition:** After initial discovery, geophysics contributes to understanding the shape, size, attitude, and continuity of the mineralized zone. High-resolution 2D and 3D surveys (e.g., 3D IP, detailed seismic, borehole geophysics) are vital for:
  - Creating detailed 3D models of the ore body and surrounding geology.
  - Estimating the volume and potential grade of the resource by integrating geophysical properties with geological and assay data.

  - Characterizing the spatial variability of the deposit, which is crucial for accurate resource estimation and economic viability assessments.

- **Mine Planning and Design:** Geophysics provides critical information for safe and efficient mine operations. This involves:
  - **Geotechnical Assessment:** Mapping bedrock depth, identifying weak zones, faults, and fracture networks (e.g., using seismic refraction, GPR, resistivity) that could impact slope stability in open pits or roof/wall stability in underground mines.
  - **Hydrogeological Assessment:** Locating aquifers, mapping groundwater pathways, and identifying potential water inflow zones (e.g., using resistivity, EM, GPR) that are critical for dewatering strategies.
  - **Overburden Characterization:** Determining thickness and properties of overburden for excavation planning and waste management.
- **Operational Monitoring:** During active mining, geophysics can be used for ongoing monitoring and optimization:
  - **Ore Body Tracking:** Geophysical surveys can help delineate the precise boundaries of the ore body as mining progresses, ensuring efficient extraction and minimizing dilution.
  - **Ground Stability Monitoring:** Microseismic monitoring arrays are routinely used in deep underground mines to detect small seismic events indicating rock mass stress and potential rockbursts, contributing significantly to mine safety (Young and Power, 2004). Similarly, InSAR (Interferometric Synthetic Aperture Radar) can monitor surface deformation in open pit mines or around tailings dams.
  - **Water Management:** Monitoring changes in groundwater levels and flow patterns.
- **Environmental Management and Post-Closure:** Geophysics plays a crucial role in mitigating environmental impacts and managing mine sites after operations cease:
  - **Tailings Dam Characterization:** Assessing the integrity of tailings storage facilities, detecting seepage, and mapping water saturation levels.
  - **Acid Mine Drainage (AMD) Detection:** Identifying pathways for acidic water migration from mine waste piles or underground workings (e.g., using electrical methods).

- **Rehabilitation Monitoring:** Assessing the success of reclamation efforts, including soil compaction and water infiltration.
- **Legacy Issues:** Locating buried mine workings or contaminants from historical mining operations.

4.1.2 Fundamental Principles

The ability of geophysics to "see" into the Earth hinges on the fact that different rocks, minerals, fluids, and geological structures possess distinct physical properties. Mineral deposits are rarely isolated phenomena; they form within specific geological contexts and are often associated with altered host rocks, structural features, or unique mineral assemblages that create measurable contrasts in these properties.

Key physical properties exploited in mining geophysics include:

- **Density:** The mass per unit volume. Massive sulfide deposits (e.g., rich in pyrite, chalcopyrite) or chromite often have higher densities than surrounding host rocks, creating positive gravity anomalies. Conversely, highly altered zones or porous weathered rocks might have lower densities.
- **Magnetic Susceptibility:** A measure of how easily a material can be magnetized. Minerals like magnetite and pyrrhotite are highly magnetic and can produce strong magnetic anomalies. Many other common rock-forming minerals are non-magnetic. This contrast is fundamental for magnetic surveys, allowing the detection of iron ores or mapping magnetic alteration halos (e.g., magnetite destruction or addition around a deposit).
- **Electrical Conductivity/Resistivity:** The ability of a material to conduct or resist an electrical current. Massive sulfide deposits, graphite, saline groundwater, and some clay-rich alteration zones are highly conductive (low resistivity). Dry, fresh igneous rocks and barren quartzites are generally highly resistive.
- **Chargeability:** The ability of a material to store an electrical charge, often related to electrochemical reactions at the interface between metallic (electronically conductive) minerals and ionic solutions (porous rocks with groundwater). Disseminated metallic sulfides and graphite are key sources of IP anomalies.
- **Seismic Velocity:** The speed at which seismic (acoustic) waves travel through a material. This is influenced by rock stiffness, density, and porosity. Hard, unfractured rocks have

higher velocities than softer, fractured, or weathered materials. Major geological contacts, faults, and specific ore bodies can create seismic reflections or refractions.

- **Radioactivity:** The emission of gamma rays from naturally decaying radioactive isotopes (primarily Potassium-40, Uranium-238, Thorium-232). Uranium and thorium deposits are directly detected, while potassium enrichment can indicate certain alteration styles associated with gold or porphyry deposits.

Geophysical surveys are designed to measure the spatial variations (anomalies) in the Earth's natural fields (gravity, magnetic) or in fields induced by artificial sources (electrical, electromagnetic, seismic). These anomalies are then processed and interpreted to infer the distribution of the underlying physical properties and, by extension, the geology and potential mineralization. It's important to distinguish between **direct detection** (where the ore body itself has a unique geophysical signature, e.g., massive sulfide detected by EM) and **indirect detection** (where geophysics maps geological structures, alteration zones, or host rocks associated with mineralization, e.g., mapping a magnetic intrusion that hosts gold veins).

4.1.3 Advantages and Limitations of Geophysical Methods

While incredibly powerful, geophysical methods are not a panacea. A clear understanding of their strengths and weaknesses is crucial for effective application.

Advantages:

- **Non-Invasive:** Geophysics provides subsurface information without disturbing the ground, making it environmentally friendly.
- **Cost-Effective for Large Areas:** Especially airborne surveys, geophysics allows rapid coverage of vast regions at a fraction of the cost of extensive drilling. This efficiency is critical in the early stages of exploration.
- **"Sees Through" Cover:** Many geophysical methods can penetrate through barren overburden (e.g., soil, glacial till, shallow sediments) that obscures underlying geology and mineralization, which is a major advantage over traditional geological mapping or geochemical sampling.

- **Reduces Drilling Risk:** By providing more informed targets, geophysics significantly increases the success rate of expensive drilling campaigns, leading to more efficient resource discovery.
- **Provides Inter-Drill Hole Information:** Geophysics offers continuous spatial data, filling the gaps between discrete drill holes and allowing for more accurate 3D geological and mineralization models.
- **Quantitative Data:** Geophysical data are measurable physical quantities, allowing for rigorous mathematical modeling and inversion to derive quantitative subsurface parameters.

Limitations and Challenges:

- **Non-Uniqueness of Interpretation:** This is the most significant challenge. Multiple subsurface geological scenarios can produce the same geophysical anomaly. For example, a magnetic anomaly could be caused by an iron ore body, a barren magnetic dyke, or a highly magnetic alteration zone. This necessitates integration with geological, geochemical, and drilling data for robust interpretation.
- **Data Acquisition Costs:** While efficient for large areas, detailed ground surveys can still be expensive, especially in remote or difficult terrain.
- **Noise Interference:** Geophysical measurements can be affected by cultural noise (e.g., power lines, fences, buried pipes) or natural noise (e.g., lightning, diurnal magnetic variations), requiring careful survey design and advanced processing.
- **Depth Penetration Limitations:** Each method has inherent limitations on how deep it can "see," depending on the physical property contrast, the size of the target, and the surrounding geological environment (e.g., highly conductive overburden can significantly reduce the depth of penetration for EM methods).
- **Requires Geological Context:** Geophysical data cannot be interpreted in isolation. A strong understanding of the regional and local geology, the expected physical properties of the target mineralization and host rocks, and the deposit model is essential for successful application.
- **Personnel Expertise:** Proper acquisition, processing, and interpretation of geophysical data require highly skilled and experienced geophysicists.

In conclusion, applied geophysics has evolved from a niche discipline into a cornerstone of modern mineral exploration and mining. By leveraging sophisticated technology and rigorous scientific principles, it enables the industry to explore more deeply, precisely, and efficiently, ultimately contributing to the sustainable discovery and extraction of the Earth's vital mineral resources. The following sections will delve into the specific methods and their detailed applications, illustrating how these tools are wielded to unlock the secrets of the subsurface.

4.2 Potential Field Methods in Mining

Potential field methods – specifically magnetic and gravity surveys – are often the initial go-to geophysical tools in mineral exploration. Their appeal lies in their ability to cover vast areas relatively quickly and cost-effectively, providing fundamental insights into the subsurface geology. These methods measure variations in naturally occurring fields (Earth's magnetic field and gravitational field) that are perturbed by differences in the physical properties of rocks and minerals. They are particularly effective for detecting large-scale geological structures and deposits that exhibit significant density or magnetic susceptibility contrasts with their surroundings.

4.2.1 Magnetic Method

The magnetic method measures variations in the Earth's magnetic field caused by differences in the magnetic susceptibility of subsurface rocks and minerals. Magnetic susceptibility is a measure of how easily a material can be magnetized when placed in an external magnetic field. While most rock-forming minerals are weakly magnetic, certain minerals, notably magnetite (Fe3O4), pyrrhotite (Fe1−xS), and to a lesser extent, some ilmenite and iron oxides, are highly magnetic. This contrast forms the basis of magnetic surveys.

4.2.1.1 Principle

The Earth's magnetic field is a complex vector field generated primarily by convection currents in its outer core (the main field). This main field interacts with susceptible rocks in the crust, inducing secondary magnetic fields. A magnetometer measures the sum of these fields, typically focusing on the total magnetic intensity (TMI). Anomalies in the TMI map reveal areas where highly magnetic rocks or minerals are present, creating a local perturbation of the Earth's background magnetic field. The shape, amplitude, and wavelength of these anomalies depend on the geometry, depth, and magnetic properties of the anomalous body, as well as the inclination of the Earth's main magnetic field (Blakely, 1995).

4.2.1.2 Instrumentation

Modern magnetometers are highly sensitive instruments capable of detecting minute changes in the magnetic field.

- **Proton Precession Magnetometers (PPM):** These are widely used for ground surveys. They measure the total magnetic field intensity based on the precession frequency of protons in a hydrogen-rich fluid within a sensor. They are robust, relatively inexpensive, and provide absolute measurements.
- **Alkali-Vapor (Cesium or Potassium) Magnetometers:** These are optically pumped magnetometers, typically used in airborne surveys due to their extremely high sensitivity and rapid sampling rates. They measure the scalar magnitude of the total magnetic field.
- **Fluxgate Magnetometers:** These measure the magnetic field component along a specific axis and are often used for borehole surveys or in gradient configurations.

4.2.1.3 Data Acquisition

Magnetic surveys can be conducted on various platforms, each offering distinct advantages:

- **Ground Magnetics:**
  - **Methodology:** Involves systematically traversing an area on foot, carrying a magnetometer. Sensors can be mounted on poles or towed behind a person (walking mag).
  - **Resolution:** Provides very high-resolution data for detailed mapping of shallow targets or precise delineation of anomalies identified in regional surveys.
  - **Applications:** Ideal for target generation, mapping shallow dykes, faults, small intrusions, buried infrastructure, or for archaeological investigations within a known prospect area. It allows for detailed follow-up of airborne anomalies.
  - **Challenges:** Slow, labor-intensive, and sensitive to cultural noise (e.g., power lines, fences, vehicles) which can produce significant interfering anomalies.

- **Airborne Magnetics (Aeromagnetics):**
    - **Methodology:** Sensors are typically mounted in fixed-wing aircraft or helicopters, flying along systematic survey lines at a constant terrain clearance.
    - **Resolution & Coverage:** Offers rapid, cost-effective coverage of vast areas, making it the workhorse for regional reconnaissance (e.g., hundreds to thousands of line kilometers per day). Resolution is lower than ground surveys but sufficient for identifying large-scale features and initial targets.
    - **Types:** Fixed-wing for large, relatively flat areas; helicopter-borne for rugged terrain or where lower, slower flight is required for improved resolution.
    - **Advantages:** Less affected by cultural noise due to higher altitude, provides consistent data quality over large regions. Often flown in conjunction with radiometrics.
    - **Applications:** The primary tool for initial mineral exploration to identify favorable geological settings and potential mineral occurrences (Reid et al., 1990).
- **Drone Magnetics (UAV-Magnetics):**
    - **Methodology:** Magnetometers are mounted on Unmanned Aerial Vehicles (UAVs).
    - **Resolution & Flexibility:** Offers a bridge between ground and manned airborne surveys. Provides higher resolution than manned aircraft at a lower cost than ground surveys, particularly in rugged or difficult-to-access terrain. Can fly lower and slower than conventional aircraft, enhancing resolution.
    - **Applications:** Detailed target generation, mapping linear features, and environmental applications like UXO detection.
    - **Challenges:** Limited payload capacity, flight endurance, and regulatory restrictions.

4.2.1.4 Applications in Mining

The magnetic method is exceptionally versatile in mineral exploration and mining:

- **Direct Detection:**
    - **Iron Ore Deposits:** Directly maps banded iron formations (BIFs) and other magnetite-rich iron ore bodies, which are typically highly magnetic.

  - **Nickel Sulfide Deposits:** Some nickel sulfide deposits are associated with pyrrhotite, which can be highly magnetic, making them direct targets.
- **Indirect Detection (Structural and Lithological Mapping):** This is where the magnetic method truly shines, as many ore bodies themselves are non-magnetic but are associated with magnetic host rocks, alteration halos, or structural controls.
  - **Delineating Geological Contacts:** Magnetic contrasts clearly define boundaries between different rock units (e.g., mafic vs. felsic intrusions, sedimentary vs. volcanic rocks) that have varying magnetic susceptibilities.
  - **Mapping Faults and Folds:** Structural features often juxtapose rocks of different magnetic properties or represent zones of alteration, leading to distinctive linear magnetic anomalies or disruptions in magnetic patterns. These structures can be key fluid pathways for mineralization.
  - **Identifying Intrusions:** Many intrusive rocks (e.g., gabbros, granites, kimberlites, porphyries) have characteristic magnetic signatures, making them identifiable targets, especially if they are host rocks or source rocks for mineralization (Macnae, 1995). For example, kimberlite pipes (primary source of diamonds) often exhibit subtle circular magnetic anomalies due to their magnetic mineral content (e.g., ilmenite, magnetite) and distinct geometry.
  - **Mapping Alteration Zones:** Some alteration processes can either create or destroy magnetite. For instance, potassic alteration in porphyry copper deposits can enhance magnetic signatures, while phyllic or sericitic alteration can destroy magnetite, creating a magnetic "low" or "hole" within a generally magnetic background.
- **Mine-site Applications:**
  - Identifying buried pipelines, cables, and other infrastructure.
  - Locating unexploded ordnance (UXO) in historical mining areas.
  - Mapping waste dumps and tailings.

4.2.1.5 Data Processing and Interpretation

Raw magnetic data require several processing steps before interpretation:

- **Diurnal Correction:** Removal of natural daily variations in the Earth's magnetic field, typically by using a stationary base station magnetometer.
- **International Geomagnetic Reference Field (IGRF) Removal:** Subtraction of the regional, long-wavelength component of the Earth's main magnetic field to emphasize local anomalies.
- **Levelling:** Adjusting for small inconsistencies between survey lines.
- **Transforms and Enhancements:** Application of various filters and transforms to enhance subtle features or aid interpretation:
    - **Reduction to the Pole (RTP):** A mathematical transformation that re-positions magnetic anomalies to be directly over their causative bodies, simplifying interpretation, especially at high magnetic latitudes.
    - **Analytical Signal:** Produces an anomaly that is always positive and centered over the causative body, regardless of the Earth's field inclination.
    - **First Vertical Derivative (1VD) / Total Horizontal Derivative (THD):** Enhances shallow features and sharpens contacts, highlighting edges of magnetic bodies.
    - **Tilt Angle Derivative (TDR):** Highlights edges and can help estimate depths.
- **Modeling and Inversion:**
    - **Forward Modeling:** Building a geological model and calculating its predicted magnetic response for comparison with observed data.
    - **3D Inversion:** A more advanced technique that directly calculates the most probable 3D distribution of magnetic susceptibility that explains the observed magnetic anomalies. This provides detailed 3D models of geological bodies.

4.2.1.6 Case Studies

- **Iron Ore Belts (e.g., Hamersley Basin, Australia; Lake Superior Region, USA/Canada):** Regional aeromagnetic surveys are fundamental for mapping vast extensions of magnetite-rich banded iron formations, which are the primary source of iron ore. The high magnetic susceptibility contrast makes these deposits easily detectable over regional scales.

- **Kimberlite Exploration (e.g., Lac de Gras, Canada; Botswana):** Many kimberlite pipes contain sufficient amounts of magnetic minerals (e.g., ilmenite, magnetic picroilmenite, magnetite) to produce subtle but distinct circular or oval magnetic anomalies. Detailed airborne and ground magnetic surveys are a cornerstone of diamond exploration.
- **Porphyry Copper Systems (e.g., Andean Copper Belt):** Magnetic surveys help to map the large intrusive bodies associated with these deposits and identify key alteration zones. For example, some porphyry deposits are characterized by a central magnetic high due to potassic alteration (magnetite addition), often surrounded by a magnetic low representing sericitic alteration (magnetite destruction).

4.2.2 Gravity Method

The **gravity method** measures small variations in the Earth's gravitational field caused by differences in the **density** of subsurface rocks and minerals. The gravitational attraction of a body is directly proportional to its mass, so variations in density create measurable changes in the local gravity field.

4.2.2.1 Principle

The Earth's gravitational field is primarily controlled by its mass distribution. Local variations in rock density create positive gravity anomalies (if the rock is denser than its surroundings) or negative gravity anomalies (if it's less dense). For example, a massive sulfide body (often very dense) might create a localized gravity high, while a low-density salt dome would create a gravity low. Gravity measurements are highly sensitive to topography and elevation, necessitating meticulous corrections.

4.2.2.2 Instrumentation

- **Relative Gravimeters (e.g., LaCoste & Romberg, Scintrex Autograv):** These are extremely sensitive spring-based instruments that measure differences in gravity between survey stations. They are the standard for high-precision ground gravity surveys.
- **Absolute Gravimeters:** Measure the absolute acceleration due to gravity (e.g., using free-fall methods). Primarily used for calibrating relative gravimeters and for precise geodetic applications.

- **Airborne Gravimeters:** Specialized gyrostabilized platforms equipped with highly sensitive accelerometers capable of measuring gravity from moving aircraft.
- **Marine Gravimeters:** Designed for use on ships or submarines.

4.2.2.3 Data Acquisition

- **Ground Gravity:**
  - **Methodology:** Involves measuring gravity at discrete stations. Each station requires precise GPS coordinates (latitude, longitude, and especially elevation) as gravity varies significantly with altitude.
  - **Resolution:** Provides high-resolution data for detailed mapping of specific targets.
  - **Applications:** Ideal for target delineation, especially for dense ore bodies. Often deployed after magnetic surveys have identified broad areas of interest.
  - **Challenges:** Labor-intensive, slow, and highly sensitive to terrain, requiring accurate topographic surveys.
- **Airborne Gravity / Gravity Gradiometry:**
  - **Methodology:** Measures gravity or the rate of change of gravity (gravity gradient) from aircraft. Gravity gradiometers measure the differential gravity field, which helps to separate anomalies from background noise and improve resolution.
  - **Resolution & Coverage:** Provides rapid regional coverage, useful for mapping large-scale density structures. Gravity gradiometry offers higher resolution than conventional airborne gravity.
  - **Applications:** Regional geological mapping, identification of large-scale structures (e.g., basins, uplifts), and reconnaissance for dense ore bodies or low-density diapirs.
- **Borehole Gravity:**
  - **Methodology:** Specialized gravimeters are lowered into drill holes to measure density variations around the borehole, providing detailed information that extends beyond the immediate borehole wall.

4.2.2.4 Applications in Mining

The gravity method is effective for targets with significant density contrasts:

- **Direct Detection:**
    - **Massive Sulfide Deposits:** Many massive sulfide deposits (e.g., Volcanogenic Massive Sulfides - VMS, Ni-Cu deposits) are characterized by high densities (e.g., 4.0-5.0 g/cm3) compared to typical host rocks (e.g., 2.7 g/cm3), making them excellent gravity targets (Dentith and Mudge, 2014).
    - **Chromite Deposits:** Chromite is very dense and often forms localized gravity highs.
    - **Iron Ore (Hematite):** While magnetite-rich iron ores are magnetic, hematite-rich iron ores are often non-magnetic but still very dense, making gravity a primary tool.
- **Indirect Detection:**
    - **Large-Scale Geological Mapping:** Delineating major structural features such as sedimentary basins (often lower density), granitic intrusions (variable density), and deep-seated fault zones.
    - **Salt Domes:** Used extensively in petroleum exploration, but relevant in mining if salt domes are associated with specific mineral deposits or present geotechnical challenges. Salt domes are typically low density, creating negative gravity anomalies.
    - **Alteration Zones:** Some alteration processes can lead to significant density changes (e.g., silicification, propylitic alteration).
    - **Density Mapping for Geotechnical Applications:** Gravity data can provide insights into rock mass density, which is related to rock strength and stability for mine planning and design.

4.2.2.5 Data Processing and Interpretation

Gravity data require rigorous corrections to isolate anomalies caused by subsurface density variations:

- **Drift Correction:** Removing instrumental drift over time.
- **Tidal Correction:** Accounting for tidal forces from the Moon and Sun.
- **Elevation Corrections:**

  - **Free-Air Correction:** Accounts for the decrease in gravity with increasing elevation.
  - **Bouguer Correction:** Accounts for the gravitational attraction of the rock mass between the station elevation and a reference datum (e.g., sea level). This is crucial for isolating subsurface anomalies.
- **Terrain Correction:** Removes the gravitational effect of local topography above and below the station. This is particularly important in rugged terrain.
- **Regional-Residual Separation:** Isolating smaller, localized anomalies (residual) from larger, long-wavelength regional trends (regional) which are often caused by deep crustal features.
- **Modeling and Inversion:** Similar to magnetics, 2D/3D forward modeling and inversion techniques are used to calculate the probable density distribution that explains the observed gravity anomalies.

4.2.2.6 Case Studies

- **Volcanogenic Massive Sulfide (VMS) Deposits (e.g., Flin Flon Greenstone Belt, Canada):** Many VMS deposits are characterized by dense sulfide minerals (pyrite, chalcopyrite, sphalerite, galena) concentrated into massive lenses. Detailed ground gravity surveys are highly effective at directly detecting these dense bodies, even when they are not conductive or magnetic.
- **Pothole Platinum Deposits (e.g., Bushveld Complex, South Africa):** Although the platinum-group elements themselves don't have strong geophysical signatures, the associated chromitite layers in the Bushveld Complex are dense and can be mapped using high-resolution gravity.
- **Kimberlite Pipe Delineation:** While magnetics is often the primary tool, some kimberlite pipes can also exhibit a subtle gravity signature (either positive or negative depending on their infill and alteration), aiding in their complete delineation.

In summary, magnetic and gravity methods form the cornerstone of regional and detailed mineral exploration. Their ability to efficiently map large areas and detect fundamental physical property contrasts makes them invaluable initial filters, guiding more expensive and higher-resolution surveys. However, due to the non-uniqueness of potential field data, their interpretation is most effective when integrated with geological knowledge and other geophysical datasets.

4.3 Electrical and Electromagnetic Methods in Mining

Electrical and Electromagnetic (EM) methods are indispensable tools in mineral exploration, particularly for detecting deposits that exhibit significant electrical conductivity or chargeability contrasts with their host rocks. These methods are highly sensitive to the presence of metallic minerals (especially sulfides and graphite), saline fluids, and certain clay minerals. They offer a more direct means of detecting many ore bodies compared to potential field methods, which often rely on indirect signatures.

4.3.1 Electrical Resistivity and Induced Polarization (IP)

Electrical resistivity and Induced Polarization (IP) are closely related methods that involve injecting an electric current into the ground and measuring the resulting voltage responses. They are often conducted simultaneously, as the same field setup can collect data for both.

4.3.1.1 Principle

- **Electrical Resistivity:**
    - **Concept:** Resistivity (ρ) is a fundamental electrical property of a material, representing its opposition to the flow of electric current. Its reciprocal is conductivity ($\sigma=1/\rho$).
    - **Mechanism:** In geological materials, electrical current primarily flows through electrolytic conduction via ions in pore fluids, and to a lesser extent, through electronic conduction via metallic minerals.
    - **Sensitivity:** Resistivity is highly sensitive to:
        - **Fluid Content:** Porosity and saturation of the rock.
        - **Fluid Salinity:** Saline groundwater significantly reduces resistivity.
        - **Clay Content:** Conductive clay minerals reduce resistivity.
        - **Metallic Minerals:** Electronically conductive minerals like massive sulfides (e.g., pyrite, chalcopyrite, pyrrhotite, galena), graphite, and some oxides dramatically reduce resistivity (increase conductivity).

  - **Measurement:** A direct current (DC) or low-frequency alternating current (AC) is injected into the ground through two current electrodes (A and B), and the resulting potential difference (ΔV) is measured between two potential electrodes (M and N) (Loke, 2000). The "apparent resistivity" is calculated from Ohm's Law and a geometric factor that depends on the electrode configuration.
- **Induced Polarization (IP):**
  - **Concept:** IP measures the chargeability of the subsurface, which is the ability of a material to temporarily store an electrical charge when current is applied, and then release it after the current is turned off. It is essentially a measure of the subsurface's capacitive properties.
  - **Mechanism:** The primary mechanisms for IP are:
    - **Electrode Polarization (Overvoltage):** Occurs at the interface between electronically conductive minerals (like metallic sulfides or graphite) and ionically conductive pore fluids. When current flows, ions accumulate around these mineral grains, creating a charge separation that acts like a tiny battery.
    - **Membrane Polarization:** Occurs in clay-rich rocks where fine pores and clay minerals impede ion movement, leading to charge build-up.
  - **Measurement:** IP can be measured in the **time domain** (measuring the voltage decay after current shut-off) or **frequency domain** (measuring the phase shift or amplitude reduction of the voltage response at different frequencies) (Sumner, 1976). The "chargeability" is the measured parameter, often expressed in millivolts/volt (mV/V) or milliseconds (ms).
  - **Sensitivity:** Highly sensitive to disseminated metallic sulfides (even at low concentrations), graphite, and some clay minerals.

4.3.1.2 Instrumentation

- **Resistivity/IP Meters:** Modern systems combine both functions, acting as controlled current sources and multi-channel voltage receivers.
- **Multi-Electrode Systems ("Automated Systems"):** Allow for rapid data acquisition using long cables with numerous regularly spaced electrodes connected to a central control

unit. This enables automated switching between different electrode configurations (arrays) to build up 2D and 3D subsurface images.

4.3.1.3 Data Acquisition

The choice of electrode array configuration (e.g., Wenner, Schlumberger, dipole-dipole, pole-dipole) influences the depth of investigation, resolution, and sensitivity to different geological targets.

- **2D/3D Electrical Resistivity Tomography (ERT) and IP:**
  - **Methodology:** Electrodes are laid out along lines (for 2D) or in grids (for 3D). Automated switching allows for efficient data collection with multiple current and potential electrode pairs.
  - **Imaging:** By systematically varying the spacing and configuration of current and potential electrodes, measurements are made at different "depths of investigation," allowing for the creation of resistivity and chargeability cross-sections (tomograms) or 3D volumes.
  - **Applications:** Ideal for detailed mapping of shallow to moderate-depth targets, identifying complex geological structures, and delineating ore body geometries.
- **Borehole IP/Resistivity:**
  - **Methodology:** Electrodes are deployed within a drill hole or between multiple drill holes (cross-hole tomography).
  - **Resolution:** Provides very high-resolution, localized information about the rock properties directly adjacent to or between boreholes.
  - **Applications:** Detailed characterization of known ore bodies, identifying subtle extensions, and providing ground truth for surface geophysical data.

4.3.1.4 Applications in Mining

IP and resistivity are powerful tools for detecting a wide range of mineral deposit types:

- **Direct Detection:**
  - **Disseminated Sulfide Deposits:** The primary application for IP. Many large tonnage, low-grade deposits (e.g., porphyry copper, epithermal gold, sediment-hosted massive sulfides) have disseminated sulfides (pyrite, chalcopyrite) that

produce strong IP anomalies, even if their bulk conductivity is not exceptionally high (Sumner, 1976).

- **Graphite Deposits:** Graphite is highly conductive and chargeable, making it a strong target for both resistivity (low) and IP (high) methods.
- **Massive Sulfide Halos:** While massive sulfides are primarily EM targets, their alteration halos often contain disseminated sulfides that produce IP anomalies, guiding exploration towards the core of the deposit.

- **Indirect Detection:**
    - **Mapping Alteration Zones:** Many types of hydrothermal alteration (e.g., sericitic, argillic) involve the formation of clay minerals, which can increase conductivity (lower resistivity). Pyrite enrichment in alteration zones (pyrite shells around porphyries) produces strong IP anomalies.
    - **Structural Mapping:** Faults and shear zones can be conductive (due to fluid flow or clay gouge) or resistive (if silicified), allowing for their delineation.
    - **Overburden Characterization:** Mapping thickness and electrical properties of overburden (e.g., saprolite, glacial till, lateritic cover), which is crucial for distinguishing bedrock anomalies from superficial effects.
- **Mine-site Applications:**
    - **Acid Mine Drainage (AMD) Pathways:** AMD plumes typically have low pH and high dissolved mineral content, resulting in high conductivity (low resistivity) signatures, which can be mapped to track environmental impacts.
    - **Seepage from Tailings Dams:** Mapping conductive leachate plumes.
    - **Groundwater Investigations:** Locating aquifers, mapping water quality (salinity), and tracing contaminant plumes.

4.3.1.5 Data Processing and Interpretation

- **Data Quality Control:** Removing noise, ensuring good electrode contact.
- **2D/3D Inversion:** Raw apparent resistivity and chargeability data are converted into true subsurface resistivity and chargeability distributions using sophisticated numerical inversion algorithms (Loke, 2000). These produce cross-sections or 3D volumes that can be directly interpreted in geological terms.

- **Integration:** IP and resistivity data are typically interpreted together, as they provide complementary information. For example, a low resistivity anomaly could be due to conductive sulfides or saline water, but if it's coincident with a strong IP anomaly, it strongly suggests the presence of sulfides.

4.3.1.6 Case Studies

- **Porphyry Copper Deposits (e.g., Andes Copper Belt, Southwestern US):** Large-scale 3D IP surveys are routinely used to map the extensive disseminated pyrite halos that commonly surround and extend beyond the economic copper mineralization, guiding deep drilling (e.g., exploration at Resolution Copper, Arizona).
- **VMS Deposits (e.g., Iberian Pyrite Belt, Spain/Portugal):** IP surveys are effective for detecting the disseminated sulfide feeder zones and alteration pipes that underlie the massive sulfide lenses.
- **Gold Deposits in Greenstone Belts (e.g., Abitibi Greenstone Belt, Canada):** Many gold deposits are associated with disseminated pyrite in alteration zones within shear zones or specific lithologies, making IP a key exploration tool.

4.3.2 Electromagnetic (EM) Methods

Electromagnetic (EM) methods involve the transmission of a time-varying primary electromagnetic field into the ground, which induces eddy currents in conductive subsurface bodies. These eddy currents, in turn, generate a secondary electromagnetic field, which is measured by a receiver. The characteristics of the secondary field (amplitude, phase, decay rate) provide information about the subsurface conductivity distribution.

4.3.2.1 Principle

- **Electromagnetic Induction:** According to Faraday's Law of Induction, a changing magnetic field induces an electric current in a conductor. The primary EM field from the transmitter induces eddy currents in conductive rocks or ore bodies.
- **Secondary Field Generation:** These induced eddy currents generate their own secondary magnetic field.

- **Measurement:** The receiver measures the sum of the primary and secondary fields. The amplitude and phase (in frequency domain EM) or the decay rate (in time domain EM) of the secondary field are analyzed to infer the conductivity of the subsurface (Nabighian and Macnae, 2005).
- **Skin Depth:** The depth to which an EM wave penetrates the earth is inversely proportional to frequency (in FDEM) and conductivity. Lower frequencies (FDEM) or later times (TDEM) penetrate deeper.

4.3.2.2 Instrumentation and Types of EM Systems

EM systems are broadly categorized by whether they operate in the frequency domain or time domain:

- **Frequency-Domain EM (FDEM):**
  - **Principle:** The transmitter emits a continuous wave at one or more specific frequencies. The receiver measures the amplitude and phase shift of the secondary field relative to the primary field.
  - **Instruments:** VLF-EM (Very Low Frequency), Slingram (Horizontal Loop EM - HLEM), Ground Conductivity Meters (e.g., EM31, EM34).
  - **Applications:** VLF-EM is good for mapping conductive linear features (faults, shear zones, conductive dykes); HLEM and GCMs are used for shallow mapping of conductors (e.g., massive sulfides, shallow groundwater, contaminant plumes).
  - **Limitations:** Limited depth penetration, sensitive to conductive overburden, and challenges in resolving closely spaced conductors.
- **Time-Domain EM (TDEM) / Transient EM (TEM):**
  - **Principle:** A current is transmitted into a large loop, creating a primary magnetic field. The current is then abruptly turned off, causing the primary field to collapse. This collapsing field induces eddy currents in any conductors, which then decay over time. The receiver measures the time-decaying secondary magnetic field during the "off-time."
  - **Instruments:** PROTEM, TEM-FAST, various commercial TDEM systems.
  - **Applications:** Considered the most powerful ground-based EM method for direct detection of discrete, often deeper, massive sulfide bodies (Nabighian and Macnae,

2005). The shape of the decay curve provides information about the conductor's size, depth, and conductivity.
    - **Advantages:** Less affected by conductive overburden than FDEM, better at resolving discrete conductors.
- **Magnetotellurics (MT) / Audio-Magnetotellurics (AMT):**
    - **Principle:** Utilizes naturally occurring electromagnetic fields generated by distant thunderstorms and solar wind interactions. It measures orthogonal components of the natural electric and magnetic fields at the Earth's surface to infer subsurface resistivity variations.
    - **Frequency Range:** AMT covers higher frequencies (Hz to kHz) for shallower depths, while MT covers lower frequencies (mHz to Hz) for deeper crustal imaging.
    - **Applications:** Regional deep crustal mapping, locating deep-seated conductive structures (e.g., graphite-rich shear zones), deep geothermal exploration, and imaging deep massive sulfide systems.
    - **Advantages:** Passive method (no artificial source), can achieve very deep penetration.
    - **Limitations:** Sensitive to cultural noise (power lines), and can be influenced by weather.

4.3.2.3 Data Acquisition

EM surveys can be conducted using various platforms:

- **Ground-based EM:**
    - **Methodology:** Involves deploying transmitting and receiving coils/loops on the ground in various configurations (e.g., fixed-loop, moving loop, coincident loop).
    - **Resolution:** High-resolution for detailed mapping of conductive targets.
    - **Applications:** Follow-up of airborne anomalies, detailed delineation of known conductors, and mapping shallow environmental targets.
- **Airborne EM (AEM):**
    - **Methodology:** Transmitters and receivers are mounted on fixed-wing aircraft or helicopters. Systems can be rigid-boom (e.g., for helicopter EM) or towed bird

systems. Modern AEM systems often operate at multiple frequencies or use time-domain measurements.

- **Resolution & Coverage:** Provides rapid, high-resolution coverage of large areas. Helicopter-borne EM (HEM) offers excellent resolution and can fly in rugged terrain; fixed-wing AEM covers vast areas efficiently.
- **Applications:** The primary tool for rapid reconnaissance and target generation for conductive ore bodies, mapping conductive overburden, and hydrogeological mapping.
- **Advantages:** Less affected by terrain variations, can penetrate through resistive cover.

- **Borehole EM (BHEM):**
  - **Methodology:** A large transmitting loop is laid on the surface, and a receiver probe is lowered into a drill hole.
  - **Resolution:** Excellent for detecting off-hole conductors (i.e., conductors missed by the drill hole itself) and determining their precise distance and direction from the borehole.
  - **Applications:** Essential for extending known mineralization, finding "look-alike" conductors along strike or down dip, and increasing drilling success rates.

4.3.2.4 Applications in Mining

EM methods are particularly effective for direct detection of highly conductive mineral deposits:

- **Direct Detection:**
  - **Massive Sulfide Deposits:** The quintessential EM target. Highly conductive massive sulfide bodies (e.g., VMS, Ni-Cu, sediment-hosted massive sulfides) produce strong EM responses due to the electronic conductivity of minerals like pyrite, pyrrhotite, chalcopyrite, and sphalerite.
  - **Graphite:** Naturally occurring graphite is highly conductive and can produce very strong EM anomalies. Distinguishing graphite from sulfides is a common challenge that requires integration with IP or drilling.

- **Indirect Detection:**
    - **Mapping Conductive Overburden:** Important for understanding the geology and for planning follow-up drilling, as thick conductive overburden can mask deeper bedrock conductors.
    - **Structural Mapping:** Conductive fault zones or shear zones (e.g., due to graphite or fluid flow) can be delineated.
    - **Saline Groundwater/Intrusion:** Mapping saline water pathways or saltwater intrusion in coastal areas.
- **Mine-site Applications:**
    - **Mine Waste Characterization:** Mapping leachate plumes from tailings or waste rock piles.
    - **Subsurface Infrastructure:** Locating buried pipes, cables, or old mine workings.

4.3.2.5 Data Processing and Interpretation

- **Filtering and Levelling:** Removing noise and ensuring data consistency.
- **1D, 2D, and 3D Inversion:** Converting measured EM responses into subsurface conductivity models. Modern inversion techniques produce detailed 3D conductivity volumes that can be directly used for geological modeling.
- **Conductivity-Depth Imaging (CDI):** A simplified, fast inversion that estimates conductivity as a function of depth.
- **Layered Earth Inversion:** For simpler, horizontally layered models.
- **Integration:** EM data are often interpreted in conjunction with magnetics (to help distinguish magnetic sulfides from non-magnetic conductors) and IP (to identify disseminated sulfides).

4.3.2.6 Case Studies

- **Nickel-Copper Sulfide Deposits (e.g., Voisey's Bay, Canada; Raglan, Canada):** These deposits are often massive and highly conductive, making them prime targets for both ground and airborne TDEM surveys. Borehole EM is then critical for extending the known mineralization.

- **Volcanogenic Massive Sulfide (VMS) Deposits (e.g., Finlayson Lake District, Canada; Kidd Creek, Canada):** VMS deposits are characterized by massive sulfide lenses that are highly conductive and are effectively detected by AEM and ground TEM.
- **Greenfield Exploration in Under-explored Regions:** AEM surveys are frequently deployed in underexplored greenstone belts or cratonic margins to identify conductive targets that may represent massive sulfides or graphite, leading to new discoveries.

In summary, electrical and electromagnetic methods are cornerstone techniques in mining geophysics, offering direct detection capabilities for conductive and chargeable mineral deposits. Their judicious application, coupled with careful data processing and interpretation alongside geological knowledge, significantly enhances the efficiency and success rates of mineral exploration programs globally.

4.4 Seismic Methods in Mining

While traditionally more prominent in the oil and gas industry, seismic methods are increasingly vital in mining, particularly for exploring deep-seated mineral deposits, providing high-resolution geotechnical information for mine planning, and enabling real-time monitoring of ground stability. Seismic techniques rely on the generation and recording of elastic waves (acoustic energy) that propagate through the Earth and are reflected or refracted by interfaces where there are changes in rock properties.

4.4.1 Seismic Reflection

Seismic reflection is the most powerful seismic technique for imaging complex geological structures at depth, analogous to medical ultrasound but on a much larger scale.

4.4.1.1 Principle

- **Wave Generation:** An artificial seismic source (e.g., explosives, vibroseis trucks, specialized impact sources) generates elastic waves that propagate into the Earth.
- **Reflection at Interfaces:** When these waves encounter a boundary (interface) between two geological layers with different **acoustic impedances** (a product of seismic velocity and density), a portion of the wave energy is reflected back towards the surface.

- **Recording:** Geophones (ground motion sensors) placed on the surface record these reflected waves.
- **Imaging:** By measuring the travel times of the reflected waves and knowing the seismic velocities of the subsurface layers, it is possible to reconstruct the geometry and depth of the reflecting interfaces, effectively creating a "picture" of the subsurface (Sheriff and Geldart, 1995).

4.4.1.2 Acquisition Considerations in Mining

Seismic reflection in hard rock mining environments presents unique challenges compared to sedimentary basins, primarily due to:

- **High Velocities:** Hard crystalline rocks have much higher seismic velocities, leading to shorter travel times and requiring very precise timing.
- **Complex Geology:** Hard rock terrains often exhibit complex faulting, folding, and intrusive bodies, leading to scattered seismic energy and complex reflection patterns.
- **High Attenuation:** Fractured or highly altered rocks can attenuate seismic energy, limiting depth penetration.
- **Noise:** Mining activities, cultural infrastructure, and highly variable overburden can generate significant seismic noise.
- **Source and Receiver Coupling:** Achieving good coupling in rocky terrain can be difficult.
- **Sources:** Explosives (e.g., dynamite, detonating cord) are often preferred for their high energy in hard rock. Vibroseis trucks are used in more accessible areas. Specialized small-scale sources are also developed for near-surface mining applications.
- **Receivers:** Geophones are typically deployed in linear arrays (for 2D) or gridded patterns (for 3D) to maximize data density and spatial sampling.
- **Survey Geometry:** 2D seismic lines provide cross-sectional images, while 3D surveys provide a volumetric image of the subsurface, crucial for detailed mine planning.

4.4.1.3 Applications in Mining

Seismic reflection is gaining increasing importance for:

- **Deep Structural Mapping:** This is arguably its most impactful application in hard rock mining. Seismic reflection can effectively map deep-seated faults, shear zones, folds, and lithological contacts that control the localization of many mineral deposits (e.g., orogenic gold, VMS, some base metals) that might be beyond the reach of other geophysical methods (Adam et al., 2012). It helps to understand the plumbing systems that brought mineralizing fluids to their current locations.

- **Coal and Potash Exploration:** Seismic reflection is the undisputed primary method for exploring and delineating coal seams and potash evaporite layers. These deposits typically exhibit clear and consistent acoustic impedance contrasts with surrounding sediments, allowing for very detailed mapping of their thickness, continuity, and structural deformation. This is critical for resource estimation and mine design.
- **Kimberlite Pipe Imaging:** While magnetic methods are excellent for initial detection, seismic reflection can provide detailed images of the internal structure and precise geometry of kimberlite pipes at depth, aiding in resource modeling and shaft placement.
- **Hydrocarbon Potential in Mining Regions:** Many mining districts are located in sedimentary basins that also host hydrocarbon resources (e.g., coalbed methane). Seismic reflection is directly applicable to exploring these co-located energy resources.
- **Mine-Scale Imaging (Future and Emerging):** High-resolution 3D seismic surveys are being adapted for imaging the geometry of ore bodies at depth, tracing their extensions, and identifying potential extensions that may be missed by drilling. This requires very dense and precise acquisition, often using downhole sources and receivers.

4.4.1.4 Data Processing and Interpretation

Seismic reflection data processing is complex and computationally intensive:

- **Pre-processing:** Includes noise attenuation (e.g., ground roll removal, random noise suppression), static corrections (to account for variations in elevation and weathered layer thickness), and amplitude corrections.
- **Common Mid-Point (CMP) Gathering and Stacking:** Multiple source-receiver pairs share a common reflection point, allowing data to be stacked to improve signal-to-noise ratio.
- **Velocity Analysis:** Determining the seismic velocities of different layers, crucial for accurate depth conversion and imaging.

- **Migration:** A key step that collapses diffractions and moves dipping reflectors to their true subsurface positions, producing a more accurate geological image (Yilmaz, 2001).
- **Interpretation:** Involves identifying reflectors, delineating faults and geological contacts, and interpreting seismic facies (patterns of seismic reflections that relate to rock types or depositional environments).
- **Time-Lapse (4D) Seismic:** Repeated 3D seismic surveys over time to monitor dynamic changes in the subsurface, such as fluid movement in aquifers or stress changes around mine excavations.

4.4.1.5 Case Studies

- **Deep Gold Deposits (e.g., Witwatersrand Basin, South Africa; Abitibi Greenstone Belt, Canada):** Deep seismic reflection has been crucial for understanding the complex fault systems and basin architecture that host some of the world's largest gold deposits, guiding exploration for extensions at significant depths.
- **VMS Exploration (e.g., Kidd Creek Mine, Canada):** Seismic reflection has been used to image the architecture of volcanic successions and large-scale structures that control the formation of VMS deposits, even when the sulfide bodies themselves are too small to be direct seismic targets.
- **Coal Exploration (e.g., Bowen Basin, Australia; Appalachian Basin, USA):** Routine application of 2D and 3D seismic for mapping coal seams, identifying seam splits, faulting, and gas content, optimizing mine planning and resource recovery.

4.4.2 Seismic Refraction and Shear Wave (Vs) Surveys

While reflection focuses on impedance contrasts at interfaces, refraction and shear wave surveys are primarily used for mapping layer velocities and depths in the shallower subsurface, which is critical for geotechnical and engineering applications in mining.

4.4.2.1 Principle

- **Seismic Refraction:** Measures the travel times of critically refracted seismic waves that travel along high-velocity layers (e.g., bedrock) after being refracted at an interface. The arrival times at different geophones allow for the calculation of layer velocities and depths (Telford et al., 1990).

- **Shear Wave (Vs) Surveys:** Measures the velocity of shear waves (S-waves), which propagate by particle motion perpendicular to the wave direction. Unlike P-waves, S-waves do not travel through liquids and are highly sensitive to the elastic properties (rigidity) of the material. Shear wave velocity (Vs) is directly related to **rock strength** and **stiffness**.

4.4.2.2 Acquisition

- **Sources:** Hammer blows, weight drops, or small explosives are common sources for refraction and Vs surveys. For S-waves, specialized sources (e.g., hammer striking a shear plate) are used to generate horizontally polarized energy.
- **Receivers:** Standard geophones are used for refraction (P-waves), while horizontal geophones are needed for S-wave measurements.
- **Arrays:** Linear arrays of closely spaced geophones are typical.

4.4.2.3 Applications in Mining

These methods are invaluable for geotechnical characterization and shallow engineering studies:

- **Overburden Characterization:** Determining the thickness and seismic velocity (and thus strength/density) of unconsolidated overburden layers overlying bedrock. This is crucial for planning excavation and ripping operations in open-pit mines.
- **Rippability Assessment:** Based on P-wave and S-wave velocities, engineers can determine whether overburden or rock requires blasting, mechanical ripping, or can be excavated directly, significantly impacting equipment selection and operational costs.
- **Bedrock Mapping:** Delineating the irregular surface of competent bedrock beneath unconsolidated material, critical for foundation design for mine infrastructure, tailings dams, and processing plants.
- **Slope Stability Analysis:** Assessing rock mass quality, identifying weak zones, and characterizing potential slip planes within open pits and along access roads (e.g., by mapping zones of reduced Vs).

- **Tunneling and Underground Mine Design:** Providing parameters for rock mass classification, support requirements, and stability assessment for tunnels, shafts, and large underground excavations.
- **Groundwater Mapping:** Identifying saturated zones or aquifers with distinct seismic velocity characteristics.

4.4.2.4 Data Processing and Interpretation

- **Refraction:** Analysis of first-arrival times to calculate layer velocities and depths using techniques like the generalized reciprocal method (GRM).
- **Shear Wave:** Processing involves picking S-wave arrivals and calculating velocities, often using methods like MASW (Multi-channel Analysis of Surface Waves) to derive 1D or 2D Vs profiles.
- **Integration:** Results are often presented as 2D cross-sections of velocity or rippability zones, integrated with drilling and geological logs.

4.4.2.5 Case Studies

- **Open Pit Mine Design:** Extensive use of seismic refraction and MASW surveys to characterize overburden properties, define excavation boundaries, and predict ground conditions for large-scale earthmoving operations.
- **Dam and Tailings Facility Design:** Geophysical surveys (including seismic refraction and Vs) are routinely used to assess foundation conditions, identify potential leakage pathways, and monitor stability.

4.4.3 Microseismic Monitoring

Microseismic monitoring is a passive seismic method that listens to the "sound" of the Earth's subtle fracturing and deformation, providing real-time insights into the stability of rock masses.

4.4.3.1 Principle

- **Event Detection:** As rock masses deform, fracture, or slip, they release small amounts of elastic energy in the form of microseismic events (akin to very small earthquakes, typically below Magnitude 2).

- **Propagation and Recording:** These waves propagate through the rock and are recorded by a network of highly sensitive geophones or accelerometers installed in boreholes or on mine infrastructure (Young and Power, 2004).
- **Source Location and Characterization:** By precisely timing the arrival of these microseismic waves at multiple sensors, the location (hypocenter) and magnitude of the events can be determined. Analysis of wave characteristics can also provide information about the type of rock failure.

4.4.3.2 Instrumentation and Acquisition

- **Sensors:** High-frequency geophones (10 Hz to 100 Hz or higher) or accelerometers are typically deployed in arrays.
- **Deployment:** Sensors are often installed in boreholes drilled into the rock mass surrounding active mine workings or slopes. Surface arrays can also be used.
- **Continuous Monitoring:** Data acquisition systems continuously listen for events, trigger recording when an event is detected, and process the data in near real-time.

4.4.3.3 Applications in Mining

Microseismic monitoring is a critical safety and operational tool in underground mines:

- **Rockburst Prediction and Prevention:** This is the primary application. In deep, highly stressed underground mines (e.g., deep gold, base metal mines), microseismic monitoring helps to identify zones of increasing stress and fracturing that could precede larger, potentially catastrophic rockbursts (sudden, violent rock failures). By detecting patterns of microseismicity, mine operators can issue warnings, adjust mining practices, or implement support measures (Young and Power, 2004).
- **Slope Stability Monitoring (Open Pit):** Monitoring networks on open-pit slopes can detect microseismic activity associated with progressive slope failure, allowing for early warning and mitigation efforts.
- **Caving Operations Monitoring:** In block caving or sub-level caving mines, microseismic monitoring tracks the progressive collapse of the orebody and overlying rock mass, providing insights into the caving front and aiding in production management.

- **Hydraulic Fracturing Monitoring:** For unconventional mining methods (e.g., in-situ leach, or potentially for enhanced rock permeability), microseismic arrays can monitor the propagation of induced fractures.
- **Evaluation of Ground Support:** Microseismic events can indicate the effectiveness of ground support systems and pinpoint areas where additional support is needed.

4.4.3.4 Data Processing and Interpretation

- **Event Detection and Picking:** Algorithms identify events and pick the arrival times of seismic waves at each sensor.
- **Locating Events:** Advanced algorithms use travel times to triangulate the hypocenter of each microseismic event.
- **Magnitude Estimation:** Quantifying the size of the event.
- **Source Mechanism Analysis:** Inferring the type of faulting or fracturing (e.g., shear, tensile failure).
- **Spatial and Temporal Analysis:** Mapping the distribution of microseismic events in space and observing their evolution over time to identify trends, clusters, or quiescence that may precede larger events.
- **Integration:** Results are integrated with geological models, stress measurements, and mine design plans.

4.4.3.5 Case Studies

- **Deep Gold Mines (e.g., Witwatersrand Basin, South Africa):** Microseismic monitoring is routinely employed in these deep, high-stress mines to manage rockburst risk and ensure miner safety.
- **Hard Rock Mines (e.g., Sudbury, Canada; Kiruna, Sweden):** Used to monitor ground conditions around large excavations, support systems, and active mining fronts.

In conclusion, seismic methods, while historically associated with oil and gas, are increasingly indispensable in mining. Reflection seismic offers powerful imaging capabilities for deep structures, refraction and shear wave surveys provide critical geotechnical parameters, and microseismic monitoring delivers real-time insights into mine stability, collectively enhancing the safety, efficiency, and sustainability of mining operations.

4.5 Radiometric Methods in Mining

Radiometric methods in geophysics are dedicated to measuring the natural gamma radiation emitted from the Earth's surface. This radiation originates from the radioactive decay of specific isotopes of potassium (K), uranium (U), and thorium (Th) that are naturally present in rocks and soils. While primarily known for direct detection of uranium and thorium deposits, radiometric surveys also provide valuable insights into lithological mapping, alteration zones, and structural controls relevant to a broader range of mineral systems.

4.5.1 Principle

- **Radioactive Decay:** The Earth's crust contains trace amounts of naturally unstable isotopes, primarily Potassium-40 (40K), Uranium-238 (238U) and its decay products (e.g., 214Bi), and Thorium-232 (232Th) and its decay products (e.g., 208Tl). These isotopes undergo spontaneous radioactive decay, emitting various forms of radiation, including gamma rays.
- **Gamma Ray Detection:** Radiometric surveys specifically measure the energy and intensity of these gamma rays. Each of the three main radioactive elements (40K, 238U, 232Th) emits gamma rays at characteristic energy levels (e.g., 40K at 1.46 MeV, 214Bi (from U-series) at 1.76 MeV, and 208Tl (from Th-series) at 2.61 MeV).
- **Correlation to Concentration:** The intensity of the detected gamma radiation is directly proportional to the concentration of the parent radioactive isotopes in the near-surface rocks and soils (Grasty, 1987).

4.5.2 Instrumentation

The core of radiometric surveying relies on specialized detectors that can identify and quantify gamma radiation:

- **Scintillometers:**
  - **Principle:** These are basic, handheld detectors that typically use a sodium iodide (NaI(Tl)) crystal or a bismuth germanate (BGO) crystal coupled to a

photomultiplier tube. When gamma rays strike the crystal, they produce scintillations (flashes of light) which are converted into electrical pulses.

  - **Measurement:** They measure the *total count* of gamma rays above a certain energy threshold, giving a general indication of radioactivity.
  - **Applications:** Quick reconnaissance, preliminary anomaly identification, and radiation safety checks.
- **Gamma-Ray Spectrometers:**
  - **Principle:** More advanced than scintillometers, spectrometers use larger NaI(Tl) or semiconductor (e.g., high-purity germanium) detectors capable of differentiating gamma rays based on their specific energy levels. This allows them to identify and quantify the individual contributions from K, U, and Th.
  - **Measurement:** They produce a gamma-ray energy spectrum, which is then processed to derive equivalent percentages of potassium (eK%), parts per million (ppm) of equivalent uranium (eU), and ppm of equivalent thorium (eTh). The "equivalent" term is used because the measurements are based on daughter products in the decay series, assuming secular equilibrium.
  - **Applications:** Detailed mineral exploration, geological mapping, and environmental monitoring.

4.5.3 Data Acquisition

Radiometric surveys are typically conducted on various platforms to suit different scales of exploration:

- **Ground Radiometrics:**
  - **Methodology:** Involves traversing an area with handheld or vehicle-mounted scintillometers or spectrometers. Measurements are taken at discrete points or continuously along lines.
  - **Resolution:** Provides very high-resolution, detailed data for specific target areas.
  - **Applications:** Follow-up of airborne anomalies, detailed mapping of radioactive occurrences, and assessing the grade of known uranium deposits.
  - **Challenges:** Slow and labor-intensive, sensitive to surface conditions (e.g., soil moisture, vegetation cover), and potential for radon gas interference (Grasty, 1987).

- **Airborne Radiometrics (Aeroradiometrics):**
  - **Methodology:** Large-volume gamma-ray spectrometer detectors (often housed in external "pods" or within the aircraft fuselage) are mounted on fixed-wing aircraft or helicopters. Surveys are flown along systematic parallel lines at a constant terrain clearance.
  - **Resolution & Coverage:** Provides rapid, cost-effective coverage of large areas. The resolution is lower than ground surveys but excellent for regional mapping and identifying significant anomalies. Often flown concurrently with aeromagnetic surveys (often referred to as "mag-rad" surveys).
  - **Advantages:** Efficient for regional reconnaissance, less affected by localized surface noise than ground surveys, and can penetrate through light vegetation and superficial cover.
  - **Applications:** The primary tool for uranium exploration, regional geological mapping, and identifying alteration halos.
- **Borehole Radiometrics (Gamma Logging):**
  - **Methodology:** A gamma-ray spectrometer probe is lowered into a drill hole, continuously recording the gamma-ray spectrum as it moves down.
  - **Resolution:** Provides a continuous, high-resolution vertical profile of K, U, and Th concentrations within the formations intersected by the borehole.
  - **Applications:** Essential for accurately delineating radioactive ore zones (e.g., uranium roll-fronts), identifying specific lithological units, and correlating between drill holes.
  - **Advantages:** Provides direct measurement in the subsurface, less affected by surface weathering or radon gas.

4.5.4 Applications in Mining

The radiometric method has distinct applications in mineral exploration:

4.5.4.1 Direct Detection:

- **Uranium and Thorium Deposits:** This is the most direct application. High concentrations of uranium and thorium in various geological settings (e.g., sandstone-hosted, unconformity-related, igneous, vein-type deposits) produce distinct radiometric anomalies. Aeroradiometrics is the most common reconnaissance tool for these deposits.

- **Rare Earth Elements (REEs):** Many REE deposits are associated with elevated thorium and/or uranium, making radiometrics a useful pathfinder for their exploration.

4.5.4.2 Indirect Detection (Lithological and Alteration Mapping):

- **Geological Mapping:** Different rock types often have characteristic K, U, and Th signatures. For example, granites are typically enriched in K, U, and Th, while mafic rocks (basalts, gabbros) are generally poor in these elements. Radiometric surveys can effectively delineate lithological boundaries and map regional geological units (e.g., mapping granitoid intrusions, identifying volcanic successions).
- **Alteration Zone Mapping:** Many mineralizing processes involve hydrothermal alteration that can either enrich or deplete specific radioactive elements, particularly potassium.
    - **Potassic Alteration:** Associated with many gold, porphyry copper, and base metal deposits, potassic alteration often involves the introduction of potassium (e.g., into sericite, K-feldspar, biotite). This creates a distinct **potassium anomaly** (high eK%) that can serve as a valuable exploration target or vector towards mineralization (e.g., mapping phyllic or argillic alteration zones in porphyry systems).
    - **Uranium/Thorium Mobilization/Depletion:** Some alteration processes can mobilize and redistribute uranium and thorium, creating local highs or lows that indicate hydrothermal activity.
- **Structural Mapping:** Linear radiometric anomalies or sharp discontinuities can indicate the presence of faults or shear zones that juxtapose rocks of different radiometric signatures or act as conduits for hydrothermal fluids.
- **Regolith Mapping:** Radiometric data can help in characterizing the composition and thickness of surface regolith, aiding in geochemical sampling programs.

4.5.4.3 Environmental Applications:

- **Baseline Radiation Surveys:** Establishing background radiation levels prior to mining activities.
- **Environmental Monitoring:** Tracking potential radionuclide migration from mine waste or tailings.
- **Remediation Assessment:** Evaluating the effectiveness of rehabilitation efforts in reducing radiation levels.

4.5.5 Data Processing and Interpretation

Raw radiometric data require several processing steps to produce interpretable maps:

- **Background Removal:** Subtracting cosmic and atmospheric background radiation.
- **Radon Correction:** Removing the effect of radon gas (a decay product of uranium) that emanates from the ground and can interfere with airborne measurements (Grasty, 1987).
- **Stripping:** Correcting for spectral overlap (e.g., higher energy gamma rays from Th can contribute to the U and K windows).
- **Height Correction:** Adjusting for variations in aircraft altitude above the ground.
- **Conversion to Concentrations:** Converting counts per second into equivalent concentrations of K (%), U (ppm), and Th (ppm).
- **Ratio Maps:** Creating ratio maps (e.g., K/Th, U/Th, U/K) to highlight specific alteration processes or geochemical signatures that might not be apparent from individual element maps. For example, a high U/Th ratio might indicate uranium enrichment, while a low K/Th might indicate potassic depletion.
- **Ternary Plots:** Combining K, U, and Th data into a single colored image (often Red-Green-Blue, where R=U, G=K, B=Th) to visually represent their relative proportions, which is highly effective for lithological discrimination.

4.5.6 Case Studies

- **Uranium Exploration (e.g., Athabasca Basin, Canada; Eastern European Roll-Front Deposits):** Aeroradiometrics is the foundational tool for identifying large-scale uranium anomalies related to unconformity-related deposits or roll-front systems in sedimentary basins. Ground follow-up and borehole gamma logging then refine the targets.

- **Potassic Alteration Mapping for Gold and Porphyry Deposits (e.g., Archean Greenstone Belts, Canada/Australia; Porphyry Copper Belts):** Many significant gold and porphyry deposits are associated with extensive potassic alteration (e.g., sericitization, biotite alteration) that results in elevated potassium concentrations. Radiometric surveys, particularly K/Th ratio maps, can effectively delineate these alteration footprints, even under cover, guiding exploration to more prospective areas.
- **Granitoid and Lithological Mapping:** In many terrains, aeroradiometrics is invaluable for differentiating various intrusive phases, volcanic units, and sedimentary sequences based on their inherent radiometric signatures, providing a cost-effective alternative or complement to traditional geological mapping.

In conclusion, radiometric methods, particularly airborne gamma-ray spectrometry, are powerful and often underutilized tools in mineral exploration and mining. Beyond their direct application in uranium and thorium exploration, they provide critical information for geological mapping, detecting subtle alteration halos, and understanding structural controls, thereby enhancing the overall exploration strategy and contributing to more informed decision-making throughout the mining lifecycle.

4.6 Borehole Geophysics in Mining

Borehole geophysics, also known as well logging or downhole geophysics, involves lowering specialized geophysical sensors (probes or "sondes") into drill holes to continuously measure the physical properties of the rock formations and fluids encountered. While surface geophysical surveys provide valuable information about the broader subsurface, they inevitably suffer from issues of resolution, ambiguity (non-uniqueness), and depth penetration limitations. Borehole geophysics directly addresses these challenges by providing high-resolution, in-situ measurements that are crucial for detailed characterization, correlation, and validation in mineral exploration and mine development.

4.6.1 Principle

The fundamental principle of borehole geophysics is to bring the measuring instrument into direct contact with the target geological formations. As the logging probe is pulled up or lowered down a drill hole, it records various physical parameters at discrete depth intervals. Each logging tool is designed to measure a specific physical property of the surrounding rock, such as:

- **Electrical properties:** Resistivity, conductivity, induced polarization (IP).
- **Radioactive properties:** Natural gamma radiation, neutron porosity, density (gamma-gamma).
- **Acoustic/Seismic properties:** P-wave and S-wave velocities, acoustic impedance.
- **Magnetic properties:** Magnetic susceptibility.
- **Geometric properties:** Borehole deviation, diameter (caliper).

These measurements provide a continuous vertical profile of rock properties, revealing lithological changes, mineralization zones, fluid content, and structural features with a level of detail often unattainable from surface surveys or even core logging alone (Keys, 1990).

4.6.2 Instrumentation

Borehole geophysical systems consist of several key components:

- **Logging Probe (Sonde):** The actual sensor unit, which is typically a cylindrical tool designed to fit within the drill hole. Modern probes are often modular, allowing multiple sensors to be run simultaneously.
- **Logging Cable:** A multi-conductor cable that transmits power down to the probe and data up to the surface unit. It also serves as the mechanical link for lowering and raising the probe.
- **Winch System:** Used to precisely control the depth and speed of the logging probe within the borehole.
- **Surface Acquisition Unit:** A computer-based system that controls the logging operation, acquires the raw data, processes it, and displays it in real-time.

A wide array of specialized logging tools exists, each tailored to measure specific physical properties:

- **Gamma Ray Log:** Measures the natural radioactivity (K, U, Th) of formations. Excellent for lithological correlation (e.g., shales/clays are often high gamma), identifying radioactive mineralization (Uranium, Thorium), and mapping potassic alteration associated with some ore deposits.
- **Density Log (Gamma-Gamma):** Uses a gamma-ray source and detector to measure bulk density by recording attenuated gamma rays scattered back from the formation. Critical for gravity modeling, rock strength estimation, and porosity determination.
- **Resistivity/Conductivity Log:** Measures the electrical resistivity or conductivity of formations using various electrode arrays. Essential for mapping conductive ore bodies (e.g., massive sulfides), groundwater salinity, and clay content. Laterologs and induction logs are common variations.
- **Magnetic Susceptibility Log:** Measures the magnetic susceptibility of the formation. Provides a continuous, detailed magnetic profile that can be correlated with surface magnetic surveys and identify magnetic minerals.
- **Sonic/Acoustic Log:** Measures the travel time of an acoustic wave through the formation, yielding compressional (P-wave) and sometimes shear (S-wave) velocities. Used for seismic modeling, rock mechanical properties (e.g., Young's Modulus, Poisson's Ratio), and identifying fractured zones.
- **Caliper Log:** Measures the diameter of the borehole, indicating washouts, caving, or fractured zones, which is important for understanding borehole conditions and correcting other logs.
- **Optical and Acoustic Televiewers:** These tools provide high-resolution images of the borehole wall (optical in clear fluid, acoustic in any fluid), allowing for detailed structural analysis (e.g., identifying fractures, bedding planes, and their orientations) and direct visual inspection of core loss intervals.
- **Induced Polarization (IP) Log:** Measures chargeability downhole, providing highly localized detection of disseminated metallic sulfides, even where surface IP responses might be ambiguous or weak.
- **Neutron Porosity Log:** Uses a neutron source to measure hydrogen content (primarily water) in the formation, providing an estimate of porosity. Important for hydrogeological studies and some resource estimates.

4.6.3 Applications in Mining

Borehole geophysics provides multifaceted benefits across the mining lifecycle, complementing and enhancing surface surveys and drill core analysis:

- **Geological and Lithological Correlation:**
    - **Subsurface Stratigraphy:** Logging provides continuous, objective measurements that can be used to accurately delineate geological units and correlate them between widely spaced drill holes, even when core recovery is poor or geological interpretation is ambiguous.
    - **Faults and Fractures:** Logs sensitive to changes in density, resistivity, or acoustic properties (like acoustic televiewers) can precisely locate and characterize fault zones, fracture networks, and shear zones that may control fluid flow or mineralization.
    - **Vein Detection:** Narrow veins, even if missed by core, might show up as distinct anomalies in high-resolution logs.
- **Ore Body Delineation and Grade Estimation:**
    - **Precise Contacts:** Logs can accurately define the top and bottom contacts of an ore body, which may be difficult to pick precisely from core alone due to assaying resolution or core loss.
    - **Internal Characterization:** Identifying variations within the ore body, such as different mineralization styles, grade distribution, or the presence of barren zones. For example, a gamma-ray log can precisely define uranium ore zones, or a resistivity log can map the extent of massive sulfides.
    - **Grade Estimation:** In some cases, a direct relationship between a geophysical parameter (e.g., density for iron ore, radioactivity for uranium) and ore grade can be established, allowing for continuous, in-situ grade estimation.
    - **Validation of Surface Surveys:** Borehole logs provide invaluable "ground truth" to calibrate and validate interpretations from surface geophysical surveys, helping to resolve ambiguities and refine 3D models.
- **Geotechnical Characterization:**
    - **Rock Mass Quality:** Sonic and density logs are critical for determining rock mechanical properties (e.g., Young's Modulus, Poisson's Ratio, unconfined

compressive strength), which are essential for mine design, stability analysis of open-pit slopes, and support requirements in underground excavations.

  - **Fracture Characterization:** Televiewer logs directly image fractures, allowing for their orientation, density, and aperture to be measured, which is crucial for assessing rock mass strength and permeability.
  - **Stress Analysis:** Acoustic logs can sometimes be used to infer anisotropic stress fields.

- **Hydrogeological Investigations:**
  - **Aquifer Identification:** Resistivity, neutron porosity, and gamma logs can identify porous and permeable layers that act as aquifers.
  - **Groundwater Flow:** Flowmeters can measure actual fluid movement within boreholes. Resistivity logs can differentiate between fresh and saline water.
  - **Permeability Assessment:** Logs can help infer permeability zones, guiding dewatering strategies or assessing potential for fluid migration from tailings.
- **Complementary to Core Logging:**
  - **Continuous Record:** Provides a complete record of properties, even in zones of poor core recovery (e.g., fractured ground, unconsolidated sediments).
  - **In-situ Measurement:** Logs measure rock properties in their natural, undisturbed state, unlike core samples that may experience stress relief or alteration upon removal.
  - **Cost-Effectiveness:** While drilling is expensive, logging a hole is often more cost-effective than taking continuous core over long intervals.

4.6.4 Key Logging Tools and Their Specific Contributions

- **Natural Gamma Ray Log:** Identifies clay-rich zones (high K), maps uranium/thorium mineralization, and delineates stratigraphy.
- **Density Log:** Provides bulk density for gravity models, rock mechanics (e.g., for calculating overburden weight), and porosity estimation.
- **Resistivity/Conductivity Logs:** Delineates conductive ore bodies (e.g., massive sulfides, graphite), characterizes groundwater (salinity), and identifies permeable zones.

- **Magnetic Susceptibility Log:** Pinpoints magnetic minerals, helps correlate with surface magnetic surveys, and refines geological mapping.
- **Sonic/Acoustic Log:** Measures P and S wave velocities, used for seismic processing, rock strength calculations, and identifying fractures.
- **Caliper Log:** Crucial for quality control of other logs (e.g., compensating for borehole irregularities) and identifying zones of caving or unstable ground.
- **Optical/Acoustic Televiewer:** Provides direct images of the borehole wall for detailed structural analysis (e.g., fracture orientation, dip of bedding planes), crucial for understanding geotechnical stability and ore body geometry.
- **Induced Polarization (IP) Log:** Highly sensitive to disseminated metallic sulfides, providing direct evidence of mineralization even at low concentrations, which can be critical for defining halos around economic deposits.

4.6.5 Data Processing and Interpretation

- **Quality Control:** Essential to ensure logs are free of noise, drift, and environmental effects.
- **Environmental Corrections:** Compensating for borehole conditions (diameter, fluid type, temperature).
- **Lithology and Mineralization Interpretation:** Interpreting log responses in terms of rock types, alteration zones, and mineralized intervals by comparing them to known signatures and core data.
- **Quantitative Analysis:** Deriving quantitative parameters such as porosity, density, and potentially ore grade from calibrated logs.
- **Cross-Plotting:** Analyzing relationships between different log parameters to identify specific rock types or fluid compositions.
- **Integrated Interpretation:** Combining borehole logs with geological core logs, surface geophysical data, and geochemical assays to build comprehensive 2D and 3D subsurface models. This integrated approach minimizes ambiguity and maximizes the value of all acquired data.

4.6.6 Case Studies

- **Uranium Ore Body Delineation (e.g., Athabasca Basin, Canada; Kazakhstan):** Borehole gamma logging is the absolute workhorse for defining the exact thickness and grade of uranium ore zones, guiding selective mining and reconciliation.
- **Massive Sulfide Deposit Characterization (e.g., VMS deposits, Ni-Cu deposits):** Downhole EM and resistivity logs are routinely used to extend the boundaries of massive sulfide lenses beyond the drill hole, detect off-hole conductors, and guide subsequent drill programs.
- **Geotechnical Assessment for Mine Infrastructure (e.g., Shaft Sinking, Tunnels):** Sonic, density, and televiewer logs provide critical information on rock strength, fracturing, and water-bearing zones, informing decisions on ground support, blasting design, and dewatering.
- **Coal Seam Quality Control:** Gamma-ray, density, and resistivity logs are used to determine coal seam thickness, ash content, and continuity, which is critical for mine planning and resource management in coal operations.

In essence, borehole geophysics acts as the vital link between surface exploration data and the ground truth provided by physical samples (core). It transforms discrete drill hole information into continuous, high-resolution subsurface profiles, drastically improving the accuracy and confidence of geological models, resource estimates, and geotechnical assessments throughout the mining lifecycle. Its role is becoming increasingly crucial as exploration targets move to greater depths and as mining operations become more complex and data-driven.

4.7 Integrated Geophysical Approaches and Future Trends

The complexity of modern mineral deposits and the increasing challenges of exploring under cover or at greater depths necessitate a departure from single-method approaches. Today's successful mining geophysics is characterized by the **integration of multiple geophysical methods**, combined with robust geological and geochemical data, and leveraged by powerful computational tools. This holistic approach significantly reduces exploration risk, provides more robust interpretations, and ultimately leads to more efficient resource discovery and sustainable mine operations.

4.7.1 Integrated Multi-Method Surveys

The principle behind integrated surveys is that different geophysical methods respond to different physical properties, providing complementary insights into the subsurface. No single geophysical method is a silver bullet; each has its strengths and weaknesses, and its own set of interpretational ambiguities. By combining them, the limitations of one method can be compensated for by the strengths of another, leading to a more constrained and confident geological model.

- **Synergistic Interpretation:** Instead of analyzing datasets in isolation, the power of integration lies in bringing together various maps, profiles, and 3D models to form a coherent understanding. For example:
    - A magnetic anomaly might indicate an intrusive body. If this anomaly is coincident with a gravity high, it could suggest a dense, mafic intrusion. If it also has an associated IP anomaly, it could point towards disseminated sulfides within or around that intrusion (Dentith and Mudge, 2014).
    - An EM conductor might indicate massive sulfides or graphite. If it is non-magnetic, it might favor graphite. If it is also associated with a weak IP anomaly, it could suggest barren massive pyrite, whereas a strong IP anomaly might indicate a pyrrhotite-rich, economic sulfide body.
    - Seismic reflection can map deep-seated faults and shear zones that control mineralization, while gravity or magnetics might highlight alteration zones or intrusions associated with those structures.
- **Targeting Strategy: From Regional to Local:** A common and effective exploration strategy involves a multi-stage, multi-method approach that progresses from broad reconnaissance to detailed targeting:

1. **Regional Airborne Surveys (e.g., Airborne Magnetics, Radiometrics, AEM):** These cost-effective surveys quickly identify large-scale geological domains, structural corridors, and major anomalies across vast areas. They serve as the first filter to narrow down prospective regions.

2. **Detailed Ground Surveys (e.g., IP, Ground EM, detailed Gravity/Magnetics):** Once regional "hot spots" are identified, ground-based surveys provide higher resolution data over smaller areas. These aim to resolve specific anomalies, delineate their geometry, and refine potential drill targets.
3. **Borehole Geophysics (e.g., Downhole EM, IP, Sonic, Televiewers):** After initial drilling, borehole geophysics provides critical validation, extends the understanding of known mineralization, detects off-hole conductors, and guides subsequent drilling programs with pinpoint accuracy.
4. **Integration with Geology and Geochemistry:** Throughout all stages, geophysical data must be continuously integrated with geological mapping, drill core logging, geochemical assays, and alteration studies. This iterative process allows geophysicists to refine their interpretations, build more realistic 3D geological models, and ultimately make informed decisions on where to drill.

4.7.2 3D/4D Modeling and Inversion

The advent of powerful computing has revolutionized geophysical interpretation, moving beyond 2D profiles to full 3D and even 4D (time-lapse) volumetric representations of the subsurface.

- **Sophisticated Software:** Modern geophysical interpretation relies heavily on specialized software platforms that can:
    - Import and manage vast quantities of multidisciplinary data (geophysics, geology, geochemistry, drilling).
    - Perform complex 3D forward modeling, where geologists propose a subsurface model, and the software calculates the expected geophysical response for comparison with observed data.
    - Execute **3D Inversion algorithms** (e.g., for magnetic, gravity, IP, EM data) that directly calculate the most probable 3D distribution of physical properties that explains the observed geophysical anomalies. This provides a more objective and quantitative image of the subsurface, reducing the non-uniqueness inherent in potential field data.

    - Build comprehensive **3D geological models** that integrate all available data, allowing geoscientists to visualize the ore body, host rocks, and structures in three dimensions.
- **Time-Lapse (4D) Geophysics:** This advanced technique involves repeating geophysical surveys (often 3D seismic, but also gravity, EM, or electrical methods) over the same area at different times. By comparing the datasets, subtle changes in physical properties can be detected, allowing for the monitoring of dynamic processes:
    - **Fluid Movement:** Tracking fluid flow in groundwater systems, fluid injection for enhanced recovery in oil/gas, or monitoring CO_2 sequestration plumes. In mining, this could involve monitoring dewatering efforts or the movement of acid mine drainage.
    - **Ground Deformation:** Monitoring subtle ground movements related to mining-induced subsidence, slope instability in open pits, or the progression of block caving operations.
    - **Resource Depletion/Production:** In some cases, changes in ore body density or fluid saturation can be monitored over time during production.

4.7.3 Machine Learning and Artificial Intelligence (AI) in Mining Geophysics

The increasing volume and complexity of geophysical data have made machine learning (ML) and artificial intelligence (AI) indispensable tools, transforming data processing, interpretation, and target prioritization.

- **Automated Interpretation and Feature Extraction:**
    - **Anomaly Detection:** ML algorithms can quickly scan large datasets to identify subtle anomalies that might be missed by human interpreters or conventional methods.
    - **Lithological Classification:** Training models on known rock types and their geophysical signatures (e.g., from well logs) to automatically classify lithologies in unknown areas from surface geophysical data.
    - **Fault and Structure Detection:** AI can be trained to recognize patterns indicative of faults, folds, and other structural features in seismic or potential field data, speeding up structural mapping.

    - **Image Processing:** Enhancing geophysical images for better visual interpretation, noise reduction, and artifact removal.
- **Predictive Modeling:**
    - **Ore Grade Prediction:** Developing models that predict ore grade or mineral concentration directly from geophysical attributes, particularly valuable where direct correlations exist (e.g., density for iron ore, magnetic susceptibility for magnetite content).
    - **Rock Property Prediction:** Predicting geotechnical parameters (e.g., rock strength, porosity, permeability) from geophysical measurements.
    - **Target Prioritization:** Using ML algorithms to integrate multiple geophysical, geological, and geochemical datasets and rank potential exploration targets based on their favorability for mineralization, significantly improving exploration efficiency.
- **Big Data Analytics:** AI/ML frameworks are essential for managing, processing, and extracting meaningful insights from the petabytes of data generated by modern geophysical surveys (e.g., large 3D seismic, regional airborne datasets).

4.7.4 Emerging Technologies

The field of mining geophysics is continuously evolving with rapid advancements in sensor technology and data acquisition platforms.

- **Quantum Technology Sensors:**
    - **Highly Sensitive Magnetometers/Gravimeters:** Research and development are underway for quantum magnetometers and gravimeters that promise orders-of-magnitude improvements in sensitivity. This could enable the detection of much smaller or deeper mineral targets with unprecedented precision, or even directly detect certain elements (e.g., Noble, 2021).
    - **Applications:** Revolutionary potential for detecting subtle geological features, very deep deposits, or small, high-value ore bodies that are currently undetectable.
- **Fiber Optic Sensing (DAS/DTS):**

  - **Distributed Acoustic Sensing (DAS):** Utilizes standard fiber optic cables as continuous arrays of seismic sensors. This enables real-time, high-density acoustic monitoring over long distances.
  - **Distributed Temperature Sensing (DTS):** Measures temperature continuously along the fiber.
  - **Applications in Mining:**
    - **Microseismic Monitoring:** Continuous monitoring of rock mass stability in underground mines or open pits, offering superior spatial coverage compared to discrete geophone arrays (Parker et al., 2014).
    - **Mine Infrastructure Monitoring:** Detecting ground deformation around pipelines, tunnels, and tailings dams.
    - **Fluid Flow Monitoring:** Tracking groundwater movement or fluid injection (e.g., for in-situ recovery).
    - **Geothermal Monitoring:** Real-time temperature and acoustic monitoring in geothermal wells.
- **Robotics and Autonomous Systems (UAVs/Drones, UGVs):**
  - **Unmanned Aerial Vehicles (UAVs/Drones):** Increasingly common for magnetic, radiometric, lidar, and even lightweight EM/GPR surveys. They offer flexible, high-resolution, and cost-effective data acquisition in hazardous or inaccessible terrain.
  - **Unmanned Ground Vehicles (UGVs):** Robots equipped with geophysical sensors for surveying in dangerous underground environments, pit walls, or contaminated areas, enhancing worker safety and efficiency.
  - **Autonomous Data Collection:** Future systems will be capable of fully autonomous mission planning, data acquisition, and real-time initial processing.

4.7.5 Addressing ESG (Environmental, Social, Governance) Concerns

In an era of increasing societal and regulatory scrutiny, applied geophysics is not just about finding minerals; it's also a critical tool for promoting sustainable and responsible mining practices:

- **Reducing Environmental Footprint:**

- **Optimized Drilling:** By providing precise targets, geophysics reduces the number of unnecessary drill holes, minimizing ground disturbance, waste generation, and associated costs.
- **Reduced Exploration Footprint:** Airborne and remote sensing geophysical methods cover large areas with minimal ground disturbance compared to traditional ground-based exploration methods.
- **Environmental Monitoring:** Geophysics is essential for monitoring environmental impacts (e.g., acid mine drainage plumes, groundwater contamination, stability of tailings storage facilities) both during operations and post-closure, facilitating proactive mitigation and remediation.

- **Enhancing Social License to Operate:** Demonstrating a commitment to responsible exploration and environmental stewardship through the use of non-invasive and monitoring technologies.
- **Ensuring Worker Safety:** Real-time microseismic monitoring, ground stability assessments, and remote sensing of deformation directly contribute to safer working conditions in both open-pit and underground mines.
- **Resource Efficiency:** Better delineation of ore bodies and waste rock reduces energy consumption and waste generated during mining, contributing to more efficient resource extraction.

4.7.6 Challenges and Opportunities

Despite these advancements, mining geophysics faces ongoing challenges and exciting opportunities:

- **Exploration under Cover:** Many of the easily accessible, outcropping deposits have been found. The future of mineral exploration lies in detecting deep-seated deposits under extensive cover (e.g., deep weathering, thick sedimentary basins, post-mineralization volcanics). This demands more powerful sources, more sensitive receivers, and sophisticated inversion techniques.
- **Resolution and Ambiguity:** Improving resolution to define orebody boundaries with greater precision and reducing the inherent non-uniqueness of geophysical interpretations

remain key challenges, driving the development of advanced algorithms and multi-physics inversion.

- **Cost-Effectiveness:** Balancing the need for high-quality, high-resolution data with cost constraints, especially in a cyclical commodity market. Automation and new platforms (e.g., drones) offer opportunities here.
- **Data Integration and Democratization:** Developing seamless workflows for integrating diverse datasets and making geophysical data and interpretation tools more accessible to a broader range of geoscientists.
- **Training and Expertise:** A continuing need for highly skilled geophysicists who can leverage these advanced technologies and integrate them effectively with geological knowledge.
- **New Deposit Types:** Developing or adapting geophysical methods for emerging deposit types (e.g., critical minerals in unconventional settings, deep geothermal reservoirs).

In conclusion, contemporary applied geophysics in mining is a dynamic and rapidly evolving field. It is no longer just about finding targets; it's about comprehensive subsurface characterization, risk mitigation, operational efficiency, and environmental responsibility. The ongoing integration of multi-physics data, coupled with advances in computational power, AI, and novel sensor technologies, positions geophysics at the forefront of sustainable mineral resource development for the future.

CHAPTER 5: APPLIED GEOPHYSICS IN PETROLEUM EXPLORATION

Petroleum exploration stands as one of the most significant drivers of geophysical innovation and application over the last century. The immense economic value and strategic importance of hydrocarbon resources have spurred continuous advancements in techniques designed to identify, characterize, and monitor subsurface reservoirs. This chapter will delve into the fundamental role of applied geophysics in the petroleum industry, primarily focusing on seismic methods, while also covering the complementary contributions of potential field methods and borehole geophysics.

5.1 Introduction to Petroleum Geophysics

The global economy, for decades, has been inextricably linked to the availability of reliable and affordable energy, primarily derived from hydrocarbon resources – oil and natural gas. These invaluable resources are typically found sequestered deep within the Earth's crust, often kilometres beneath the surface, making their discovery and extraction a formidable scientific and engineering challenge. This is precisely where applied geophysics steps in as an indispensable discipline.

5.1.1 What is Petroleum Geophysics?

Petroleum geophysics is the specialized application of geophysical principles and techniques aimed at understanding the Earth's subsurface for the purpose of identifying, characterizing, and monitoring geological formations that are likely to contain economic accumulations of hydrocarbons. It is a highly quantitative and data-intensive field that employs various physical measurements to infer subsurface properties, ultimately helping to reduce the significant risks and costs associated with drilling.

The primary goal of petroleum geophysics is to translate complex geological concepts into measurable physical parameters. By analyzing the way various physical fields (e.g., seismic waves, gravity, magnetism, electromagnetic energy) interact with subsurface rocks and fluids, geophysicists can:

- **Map subsurface geological structures:** Delineating faults, folds, salt diapirs, and other structural features that can form traps for hydrocarbons.

- **Characterize rock properties:** Estimating attributes such as porosity (the amount of void space in a rock), permeability (the ability of fluids to flow through a rock), lithology (rock type), and fluid content (oil, gas, water).
- **Directly or indirectly infer the presence of hydrocarbons:** Identifying anomalies that are characteristic of oil and gas reservoirs.

Essentially, petroleum geophysics provides a "subsurface imaging" capability, allowing geoscientists to "see" through layers of rock and pinpoint potential hydrocarbon reservoirs before committing to costly drilling operations.

5.1.2 The Petroleum System and Geophysical Targets

To understand where geophysics applies, it's crucial to grasp the concept of a petroleum system. A complete petroleum system requires the presence of several critical geological elements, all occurring in the right place at the right time (Magoon and Beaumont, 1999):

1. **Source Rock:** Organic-rich sedimentary rock that, when subjected to sufficient heat and pressure over geological time, generates hydrocarbons.
2. **Reservoir Rock:** A porous and permeable rock unit (e.g., sandstone, limestone, fractured basement) capable of storing and transmitting hydrocarbons.
3. **Seal Rock (Caprock):** An impermeable rock unit (e.g., shale, evaporite) that prevents hydrocarbons from migrating further upwards and escaping.
4. **Trap:** A geological configuration (structural or stratigraphic) that physically traps hydrocarbons within the reservoir rock.
5. **Migration Pathway:** The route hydrocarbons take from the source rock to the trap.
6. **Timing:** All these elements must be in place and functional at the correct geological time for accumulation to occur.

Geophysical methods, particularly seismic, directly target several of these critical elements:

- **Reservoir Characterization:** Seismic data, especially after advanced processing and inversion, can reveal zones of high porosity and permeability. Differences in seismic velocity and density, influenced by pore fluids, allow geophysicists to infer lithology and potential reservoir quality.

- **Trap Delineation:** This is a cornerstone application. Geophysics is exceptionally good at mapping both:
    - **Structural Traps:** Such as anticlines (upward folds), synclines (downward folds), faults, and salt diapirs (where salt pierces overlying layers). Seismic reflection provides high-resolution images of these features.
    - **Stratigraphic Traps:** Such as pinch-outs (where a reservoir rock thins and disappears), reefs (porous carbonate buildups), and unconformities (buried erosional surfaces). Seismic stratigraphic interpretation is key here.
- **Fluid Identification:** While not always direct, certain geophysical signatures, particularly in seismic data (e.g., Amplitude Variation with Offset – AVO, and "bright spots"), can be **Direct Hydrocarbon Indicators (DHIs)**, suggesting the presence of gas or oil within a reservoir (Rutherford and Williams, 1989).
- **Source and Seal Rock Mapping:** While these are not typically direct geophysical targets, their presence can often be inferred indirectly by mapping the stratigraphic framework in which they occur. For instance, thick, continuous shales (potential seal rocks) often have distinct seismic characteristics.

5.1.3 Role of Geophysics in the Exploration and Production (E&P) Lifecycle

Applied geophysics is not a one-off tool but an integral part of the entire E&P value chain, constantly evolving with the needs of the industry:

- **Exploration:**
    - **Reconnaissance (Basin Analysis):** At the earliest stages, regional geophysical surveys (e.g., 2D seismic, airborne gravity, magnetics) are used to understand the fundamental geological architecture of large sedimentary basins. This helps identify areas with potential for source, reservoir, and seal rocks, and broad structural trends that could host traps. The goal is to identify play types and define large, prospective areas, often hundreds or thousands of square kilometers.
    - **Prospect Generation:** Once a promising area (a "play") is identified, more detailed geophysical surveys, predominantly 3D seismic, are acquired. These surveys are designed to precisely delineate potential hydrocarbon accumulations (called "prospects"), characterize their trapping mechanisms, estimate reservoir volumes,

and identify the most favorable locations for drilling exploratory wells. This stage reduces the prospective area to a few square kilometers.

- **Appraisal:** After a discovery well confirms the presence of hydrocarbons, geophysics is crucial for appraisal. More detailed 3D seismic surveys, often incorporating specialized processing and quantitative interpretation (QI) techniques, are used to:
    - Define the full extent and geometry of the discovered reservoir.
    - Determine the distribution of reservoir properties (porosity, net pay, fluid type).
    - Identify fluid contacts (oil-water, gas-oil).
    - Assess reservoir heterogeneity and identify potential compartmentalization.
    - Optimize the placement of appraisal wells to accurately quantify the resource.
- **Development:** Once a field is deemed commercial, geophysics continues to play a vital role in optimizing its development and maximizing hydrocarbon recovery:
    - **Well Placement:** High-resolution 3D seismic data guides the precise placement of development wells (production and injection wells) to efficiently drain the reservoir and avoid unproductive zones.
    - **Reservoir Monitoring (4D Seismic):** Repeated 3D seismic surveys (4D seismic) are used to monitor changes in the reservoir over time due to production or injection (e.g., tracking water flood fronts, gas cap expansion, or pressure changes). This real-time monitoring allows for dynamic reservoir management and optimization of Enhanced Oil Recovery (EOR) strategies (Lumley, 2010).
    - **Drilling Hazard Avoidance:** Seismic data helps identify shallow gas pockets, overpressured zones, and fault systems that could pose drilling hazards.
- **Production and Beyond:**
    - **Reservoir Surveillance:** Continued 4D seismic monitoring to optimize long-term production and identify remaining undrained hydrocarbon pockets.
    - **Abandonment Planning:** Identifying subsurface infrastructure and assessing environmental impacts.
    - **New Energy Applications:** With the global energy transition, geophysical techniques developed for petroleum are now being repurposed for **Carbon Capture, Utilization, and Storage (CCUS)**, where 4D seismic is used to monitor the injection and containment of CO2 in depleted reservoirs or saline aquifers.

Similarly, geophysics is crucial for geothermal energy exploration and subsurface hydrogen storage.

5.1.4 Why Geophysics is Essential for Petroleum

The petroleum industry's reliance on geophysics stems from several fundamental reasons:

- **Deep Targets:** Hydrocarbon reservoirs are typically located at significant depths (hundreds to thousands of meters), making direct exploration (e.g., extensive drilling) economically prohibitive without prior subsurface knowledge. Geophysics provides the only practical means to image these depths.
- **Risk Mitigation:** Drilling an offshore well can cost tens or even hundreds of millions of dollars. A dry hole (a well that doesn't find economic hydrocarbons) represents a massive loss. Geophysical surveys, by mapping potential traps and identifying indicators of hydrocarbons, drastically reduce this exploration risk, guiding drill bits to the most promising locations.
- **Cost-Effectiveness:** While geophysical surveys are expensive, they are significantly more cost-effective than drilling blindly. A single 3D seismic survey can cost millions to tens of millions of dollars, but it can provide critical data over hundreds of square kilometers, informing decisions for numerous wells.
- **Subsurface Continuity:** Well logs provide discrete data points at specific locations. Geophysics, particularly seismic, provides a continuous, spatially distributed image of the subsurface between these wells, allowing for accurate interpolation, extrapolation, and the creation of comprehensive 3D geological models.
- **Non-Invasive Nature:** Geophysical methods typically collect data from the surface, minimizing environmental impact compared to extensive drilling or excavation.

In essence, applied geophysics forms the technological backbone of the modern petroleum industry. Its continuous evolution has transformed hydrocarbon exploration from a speculative endeavor into a data-driven science, enabling the discovery and efficient development of vast energy resources that power the world. The following sections will delve into the specific geophysical methods, emphasizing the dominance and sophistication of seismic techniques, while also exploring the complementary roles of other geophysical disciplines.

5.2 Seismic Methods: The Cornerstone of Petroleum Exploration

Among all the applied geophysical techniques, seismic methods stand as the undisputed cornerstone of petroleum exploration and reservoir characterization. Their unparalleled ability to provide high-resolution images of subsurface geological structures, stratigraphy, and even fluid content makes them indispensable throughout the entire exploration and production (E&P) lifecycle. The sophistication of seismic technology has grown exponentially since its inception, moving from rudimentary 2D profiles to complex 3D and 4D volumetric imaging, underpinned by advanced processing and interpretation workflows.

5.2.1 Fundamentals of Seismic Reflection

The seismic reflection method is analogous to how bats navigate using echolocation or how medical ultrasound creates images of internal organs. It relies on generating acoustic energy and recording its echoes from subsurface geological boundaries.

- **Principle:** An artificial seismic source at or near the Earth's surface generates elastic waves (also known as seismic waves) that propagate downwards into the subsurface. When these waves encounter an interface between two geological layers with contrasting acoustic properties, a portion of the wave energy is reflected back towards the surface, while the remaining energy is transmitted and continues to propagate deeper. Receivers on the surface record these reflected waves, and by precisely measuring their travel times, geophysicists can reconstruct the geometry and depth of the reflecting interfaces, effectively creating a "picture" of the subsurface.
- **Acoustic Impedance (AI):** The key physical property governing seismic reflections is acoustic impedance (Z), defined as the product of the rock's bulk density (ρ) and the seismic compressional wave (P-wave) velocity (Vp): $Z=\rho Vp$ When a seismic wave encounters a boundary where there is a change in acoustic impedance, a reflection occurs. The greater the contrast in acoustic impedance across an interface, the stronger the reflection.
- **Reflection Coefficient (R):** The amplitude of the reflected wave relative to the incident wave is quantified by the reflection coefficient. For a simple normal incidence (perpendicular to the interface) at an interface between two layers (Layer 1 and Layer 2),

the reflection coefficient is given by: R=Z2+Z1Z2−Z1 where Z1 and Z2 are the acoustic impedances of Layer 1 and Layer 2, respectively. A positive reflection coefficient indicates an increase in impedance (e.g., from shale to denser sand), while a negative coefficient indicates a decrease (e.g., from shale to less dense gas-filled sand).

- **Wave Propagation:**
  - **P-waves (Compressional Waves):** These are the primary waves used in conventional seismic reflection. Particle motion is parallel to the direction of wave propagation (like a sound wave). They can travel through solids, liquids, and gases.
  - **S-waves (Shear Waves):** Particle motion is perpendicular to the direction of wave propagation. S-waves are slower than P-waves and cannot travel through fluids. They are increasingly used in specialized seismic applications (e.g., for fracture detection, fluid discrimination) as they provide complementary information to P-waves.
  - **Other Waves:** Surface waves (e.g., Love waves, Rayleigh waves, often referred to as "ground roll") and multiple reflections (waves that reflect more than once) are typically considered noise and must be suppressed during processing.
- **Seismic Trace:** A seismic trace is a recording of the ground motion at a single receiver over time, representing the sequence of reflected waves arriving from different depths. These traces form the fundamental building blocks of a seismic survey.
- **Seismic Section:** When numerous seismic traces are displayed side-by-side, they form a seismic section, which is a 2D cross-sectional image of the subsurface. In 3D surveys, these sections can be combined to form a volumetric image. Reflections are typically displayed as peaks and troughs, representing positive and negative reflection coefficients, and their continuity, amplitude, and phase variations carry crucial geological information.

5.2.2 Seismic Data Acquisition

Seismic data acquisition involves carefully planning and executing the deployment of sources and receivers to illuminate the target subsurface adequately. The choice of acquisition system depends on the environment (land, marine, transition zone) and the specific exploration objectives.

- **Sources:** The devices used to generate the elastic waves:
  - **Land Sources:**

        - **Vibroseis Trucks:** The most common land source. These large trucks use a hydraulic vibrator plate to sweep a controlled range of frequencies into the ground (typically 10-100 Hz). This allows for precise control over the input energy and is highly repeatable.
        - **Explosives (Dynamite):** High-energy source, often used in hard-rock environments or for deep targets where maximum penetration is required. Charges are typically buried in shallow boreholes.
        - **Accelerated Weight Drop:** A heavy weight is dropped onto a baseplate, generating an impulsive seismic wave. Less energetic than vibroseis or dynamite, usually for shallow surveys.
    - **Marine Sources:**
        - **Airguns:** The predominant marine source. Arrays of airguns release highly compressed air into the water, creating an expanding bubble that generates a powerful acoustic pulse. They are towed behind a seismic vessel.
    - **Transition Zone Sources:** For areas where land meets water (e.g., estuaries, marshes), hybrid approaches are used, often employing smaller airgun arrays on shallow-draft boats or adapted land sources (e.g., "marsh buggies" for vibroseis, or borehole dynamite) in marshy terrain.
- **Receivers:** The devices that detect the returning seismic waves:
    - **Land Receivers:**
        - **Geophones:** Electromechanical sensors that convert ground motion (particle velocity) into an electrical signal. They are typically deployed in arrays to enhance the signal and suppress unwanted noise.
    - **Marine Receivers:**
        - **Hydrophones:** Pressure-sensitive transducers that convert pressure waves in water into electrical signals. They are housed within long, fluid-filled cables called **streamers** (typically 6-12 km long, containing thousands of hydrophones), which are towed behind the seismic vessel.
        - **Ocean Bottom Seismic (OBS) / Ocean Bottom Nodes (OBN):** These systems involve deploying both geophones and hydrophones (or combined sensors) directly on the seafloor.

        - **OBS:** Nodes are typically tethered to a recording vessel.
        - **OBN:** Autonomous, cable-free nodes are deployed and later retrieved.
        - **Advantages:** Provide "full wavefield" recording (both P-waves and S-waves), superior imaging in areas with complex overburden (e.g., sub-salt, sub-basalt where streamer data is poor), and can be deployed around obstacles like production platforms.
    - **Distributed Acoustic Sensing (DAS):** An emerging technology that uses standard fiber optic cables as continuous arrays of acoustic sensors. Changes in light backscatter due to strain (from seismic waves) are measured. This allows for extremely high-density spatial sampling and is increasingly used for vertical seismic profiles (VSP), microseismic monitoring, and potentially for surface seismic acquisition (Parker et al., 2014).
- **Survey Design and Geometry:** The arrangement of sources and receivers is critical for acquiring high-quality data and dictates the type of seismic image produced.
    - **2D Seismic:** Involves a single source line and a single receiver line. This provides a cross-sectional image of the subsurface along that line. 2D surveys are cost-effective for regional reconnaissance and initial basin analysis.
    - **3D Seismic:** The industry standard for detailed exploration and reservoir characterization. It involves acquiring a dense grid of source and receiver points over a defined area, resulting in a volumetric (3D) image of the subsurface. This allows geoscientists to visualize complex structures and stratigraphy in three dimensions, reducing drilling risk significantly.
        - **Land 3D:** Typically involves orthogonal survey designs (source lines perpendicular to receiver lines) to ensure uniform coverage.
        - **Marine 3D:** Multiple streamers are towed side-by-side behind the vessel, often with dual source arrays for efficiency.
        - **Wide-Azimuth (WAZ) and Full-Azimuth (FAZ) Seismic:** Advanced 3D acquisition geometries designed to collect seismic data from a wide range of source-receiver azimuths. This provides superior illumination of

complex subsurface features (e.g., sub-salt structures, fractured reservoirs) by collecting energy that reflects off surfaces at various angles.

- **4D Seismic (Time-Lapse Seismic):** This involves acquiring repeated 3D seismic surveys over the same area at different times (e.g., before production, then every few years during production). By comparing the seismic volumes, subtle changes in seismic attributes (e.g., amplitude, velocity) can be detected, which are caused by changes in fluid saturation (oil/gas depletion, water/gas injection) or pressure within the reservoir (Lumley, 2010). 4D seismic is invaluable for:
  - Monitoring fluid movement (e.g., tracking water flood fronts).
  - Identifying bypassed oil/gas.
  - Optimizing production strategies and Enhanced Oil Recovery (EOR) efforts.
  - Monitoring CO2 sequestration in depleted reservoirs or saline aquifers.

5.2.3 Seismic Data Processing

Raw seismic data, often containing significant noise, must undergo a complex and computationally intensive series of processing steps to transform it into a high-fidelity image of the subsurface. The goal is to remove noise, enhance signal, and correctly position subsurface reflectors.

- **Key Processing Steps:**
  - **Pre-processing and Quality Control (QC):** Initial steps include reviewing data quality, removal of spurious traces, and basic noise suppression (e.g., muting of ground roll).
  - **Static Corrections:** Compensating for variations in topography, elevation, and the presence of a low-velocity weathered layer near the surface. This ensures that arrival times reflect subsurface geology, not surface irregularities.
  - **Gain Recovery:** Applying amplitude corrections to compensate for geometric spreading and intrinsic attenuation of seismic energy as it travels through the Earth.
  - **Common Mid-Point (CMP) Gathering:** Traces are sorted into groups based on their common midpoint between the source and receiver. Each CMP gather contains traces reflecting from approximately the same subsurface point but at different source-receiver offsets.

- **Velocity Analysis:** A critical step that determines the average seismic velocities of the subsurface layers. This is done by analyzing how reflection times vary with offset within CMP gathers (Normal Moveout - NMO). Accurate velocity models are paramount for correct depth conversion and imaging.
- **Normal Moveout (NMO) Correction:** Applying a time correction to each trace in a CMP gather to align reflections horizontally. This is based on the velocity model.
- **Stacking:** Summing (stacking) all the NMO-corrected traces within a CMP gather. Stacking significantly enhances the signal-to-noise ratio, as coherent reflections add constructively while random noise adds destructively.
- **Deconvolution:** A process to sharpen the seismic wavelet (the basic seismic pulse) and remove reverberations (e.g., ghost reflections, short-period multiples) to improve resolution.
- **Multiple Suppression:** Attenuating seismic energy that has reflected more than once (e.g., from the seafloor and then the subsurface). Multiples can obscure primary reflections and lead to misinterpretations.
- **Migration:** The most crucial imaging step. Migration repositions dipping reflectors to their true subsurface spatial locations and collapses diffractions (energy scattered from points or edges) back to their origin. This process transforms a time section into a more accurate geological image.
    - **Time Migration:** Repositions reflectors in time.
    - **Depth Migration:** A more advanced and computationally intensive process that directly images reflectors in true depth, requiring a highly accurate velocity model. Essential for complex structures like salt bodies or sub-basalt imaging.
    - **Pre-Stack Migration:** Migration applied before stacking, utilizing the full seismic wavefield for superior imaging, especially in complex areas.
- **Amplitude Preserving Processing:** Workflows designed to preserve the true relative amplitudes of reflections, which is essential for quantitative interpretation techniques like AVO analysis.

5.2.4 Seismic Data Interpretation and Attributes

Once processed, seismic data becomes the primary tool for geological interpretation, allowing geoscientists to build comprehensive subsurface models.

- **Structural Interpretation:** The fundamental step involves identifying and mapping key horizons (reflectors representing geological boundaries), faults, folds, unconformities, and salt bodies. This delineates the subsurface trapping geometries that could contain hydrocarbons.
- **Stratigraphic Interpretation:** Involves interpreting seismic facies (patterns of seismic reflections) to understand depositional environments, differentiate between various lithologies (e.g., sands, shales, carbonates), and identify stratigraphic traps (e.g., pinch-outs, reefs).
- **Seismic Attributes:** Beyond basic amplitude and time, geophysicists extract various quantitative measurements (attributes) from the seismic data to highlight subtle geological features or directly infer rock and fluid properties. Examples include:
  - **Coherence/Similarity:** Highlights discontinuities and faults.
  - **Curvature:** Delineates subtle flexures and fracture patterns.
  - **Spectral Decomposition:** Analyzes frequency content to identify thin beds or fluid effects.
  - **Sweetness/RMS Amplitude:** Can highlight zones of higher porosity or fluid content.
  - **Amplitude Variation with Offset (AVO):** One of the most important seismic attributes for direct hydrocarbon detection. AVO analysis examines how the amplitude of a seismic reflection changes with increasing offset (distance between source and receiver), which is related to the angle of incidence at the reflector. Different fluid types (oil, gas, water) affect the elastic properties of reservoir rocks in distinct ways, leading to characteristic AVO responses that can indicate the presence of gas, particularly in clastic reservoirs (Rutherford and Williams, 1989).
  - **Seismic Inversion:** A process that transforms seismic reflection data (which represents reflection coefficients) into quantitative rock properties, typically acoustic impedance (Z), P-wave velocity (Vp), and S-wave velocity (Vs).

  - **Sparse Spike Inversion:** Estimates AI by trying to match the seismic trace with a series of spikes (reflection coefficients).
  - **Model-Based Inversion:** Builds a geological model and iteratively adjusts its properties until the synthetic seismic response matches the observed data.
  - **Elastic Inversion:** A more advanced form that inverts for multiple elastic properties (e.g., P-impedance, S-impedance, density, Vp/Vs ratio) directly from AVO data. These elastic properties are then linked to reservoir properties (porosity, fluid saturation) through **rock physics models** (Russell, 1988; Mavko et al., 2009).
  - **Full Waveform Inversion (FWI):** A revolutionary and computationally intensive technique that utilizes the entire seismic waveform (not just reflections) to iteratively build highly accurate velocity models and high-resolution images of the subsurface by minimizing the difference between observed and modeled seismic data (Virieux and Operto, 2009). FWI is crucial for imaging beneath complex overburden like salt.
- **Direct Hydrocarbon Indicators (DHIs):** These are specific seismic anomalies that provide direct evidence for the presence of hydrocarbons. While not foolproof, they significantly reduce exploration risk. Common DHIs include:
  - **Bright Spots:** Anomalously high amplitudes often associated with gas accumulations (gas-filled sands have lower acoustic impedance than water-filled sands).
  - **Dim Spots:** Anomalously low amplitudes.
  - **Flat Spots:** A horizontal reflection within a dipping structure, often representing a gas-water or oil-water contact.
  - **Amplitude Changes with Offset (AVO anomalies):** As discussed above.

5.2.5 Case Studies in Seismic Exploration

The impact of seismic methods on petroleum exploration is best illustrated through successful applications:

- **Salt Dome Imaging (e.g., Gulf of Mexico, Brazil Pre-Salt):** Complex salt diapirs create challenging seismic imaging problems due to their high velocity and irregular shapes. The development of advanced 3D seismic acquisition (e.g., wide-azimuth, long-offset) and processing (e.g., pre-stack depth migration, Full Waveform Inversion) has been critical for imaging complex sub-salt reservoirs in these prolific basins. OBN data have proven particularly effective due to their ability to record wider angles and provide full wavefield information.
- **Deepwater Exploration (e.g., Brazilian Presalt, West Africa, Gulf of Mexico):** The exploration of deepwater plays, characterized by challenging reservoirs (e.g., turbidites, fractured carbonates) and long seismic travel paths, relies heavily on high-fidelity, broadband 3D seismic. These surveys image subtle stratigraphic features and fluid contacts in challenging environments.
- **Unconventional Resources (e.g., Shale Gas/Oil in North America):** While basic seismic has always been used, the rise of unconventional plays has driven specific seismic applications:
    - **High-Resolution 3D Seismic:** Used to map subtle geological features, natural fracture networks, and "sweet spots" (zones with optimal reservoir quality and fracability) within shale formations.
    - **Microseismic Monitoring:** Crucial for monitoring hydraulic fracturing operations. Networks of geophones (often deployed in nearby wells) detect the tiny seismic events generated as the rock fractures, providing real-time maps of the stimulated reservoir volume and fracture geometry (Rutledge et al., 2004). This allows operators to optimize frac designs and well spacing.
- **Arctic Exploration:** Specialized seismic acquisition systems (e.g., ice-capable seismic vessels, over-ice vibroseis) and processing techniques are developed to operate in harsh Arctic environments, imaging beneath permafrost and sea ice to find new reserves.

In summary, seismic methods, particularly reflection seismology, are the linchpin of petroleum exploration. Their ability to image the subsurface with unprecedented detail, combined with continuous innovation in acquisition, processing, and interpretation, has made them indispensable for discovering, characterizing, and managing the world's vital hydrocarbon resources.

5.3 Potential Field Methods in Petroleum Exploration

While seismic methods provide the high-resolution images necessary for detailed prospect delineation, potential field methods – specifically magnetic and gravity surveys – play a crucial and complementary role in the early stages of petroleum exploration. They are often the initial geophysical tools deployed, providing a cost-effective and rapid means of understanding the regional geological framework of large sedimentary basins, identifying broad structural features, and guiding subsequent, more expensive seismic campaigns. These methods measure variations in naturally occurring force fields (Earth's magnetic field and gravitational field) that are perturbed by differences in the physical properties of subsurface rocks.

5.3.1 Magnetic Method

The **magnetic method** in petroleum exploration measures spatial variations in the Earth's total magnetic field caused by differences in the magnetic susceptibility of subsurface rocks. Magnetic susceptibility is a measure of how easily a material can be magnetized when placed in an external magnetic field.

5.3.1.1 Principle

The Earth's main magnetic field is generated primarily by convection currents in its molten outer core. This main field interacts with rocks in the crust, inducing secondary magnetic fields. The strength of this induced magnetization depends on the rock's magnetic susceptibility.

- **Highly Magnetic Rocks:** Generally, igneous and metamorphic rocks (especially those rich in magnetite, a common iron oxide mineral) are highly magnetic. These often form the "basement" underlying sedimentary basins.
- **Weakly Magnetic Rocks:** Most sedimentary rocks, which are the primary host for hydrocarbons, have very low magnetic susceptibility.
- **Contrast is Key:** The fundamental principle is that the large contrast in magnetic susceptibility between the relatively non-magnetic sedimentary rocks and the often highly magnetic igneous/metamorphic basement rocks creates measurable anomalies. A magnetometer detects the sum of the Earth's main field and these local induced fields.

5.3.1.2 Data Acquisition

Magnetic surveys for petroleum exploration are almost exclusively conducted from the air for efficiency over large areas.

- **Airborne Magnetics (Aeromagnetics):**
  - **Methodology:** Magnetometers are mounted in fixed-wing aircraft or helicopters, which fly along systematic parallel lines at a constant terrain clearance. A base station magnetometer is also operated on the ground to record and correct for diurnal (daily) variations in the Earth's magnetic field.
  - **Instrumentation:** Highly sensitive cesium vapor or proton precession magnetometers are typically used.
  - **Advantages:**
    - **Rapid Coverage:** Can cover vast areas (hundreds to thousands of square kilometers) quickly and cost-effectively.
    - **Regional Reconnaissance:** Provides a regional overview of subsurface geology before more detailed seismic acquisition.
    - **Relatively Inexpensive:** Significantly cheaper per line-kilometer than seismic surveys.
    - **Minimal Environmental Impact:** Non-invasive, as data is collected from the air.

5.3.1.3 Applications in Petroleum Exploration

The magnetic method, while not directly detecting hydrocarbons, provides invaluable insights into the regional structural framework of sedimentary basins, which indirectly aids in petroleum systems analysis.

- **Basement Mapping:** This is the primary application. Magnetic data helps to:
  - **Delineate Basin Boundaries:** Define the lateral extent of sedimentary basins by identifying areas where the magnetic basement (often crystalline igneous/metamorphic rocks) is shallow or outcrops, contrasting with areas where it deepens rapidly, indicating thicker sedimentary fill.
  - **Estimate Sedimentary Thickness:** By modeling the depth to magnetic basement, geophysicists can estimate the thickness of the overlying non-magnetic

sedimentary sequence, which is crucial for identifying areas with sufficient burial history for hydrocarbon generation and accumulation.
  - **Map Basement Topography:** Identify basement highs (paleo-highs) or lows that can influence subsequent sedimentation patterns and form structural traps in the overlying sediments.
- **Identification of Igneous Intrusions and Extrusions:** Magnetic anomalies can highlight buried dykes, sills, and volcanic flows within the sedimentary section. These features can:
  - **Create Traps:** Some intrusions can deform overlying sediments to form structural traps.
  - **Act as Seals:** Certain igneous rocks might act as seals.
  - **Affect Hydrocarbon Maturation:** Localized heating from intrusions can influence the maturation of source rocks.
  - **Disrupt Reservoirs:** Large igneous bodies can compartmentalize or destroy existing reservoirs.
- **Delineation of Major Faults and Structural Trends:** Deep-seated faults that cut into or offset the magnetic basement often produce linear magnetic anomalies or discontinuities in magnetic patterns. These faults can:
  - **Form Traps:** Provide seals or conduits for hydrocarbon migration.
  - **Control Basin Development:** Define the tectonic framework of a basin.
- **Salt Dome Flanks (Indirectly):** While salt itself is non-magnetic, the flanks of salt domes can sometimes be associated with subtle magnetic anomalies due to localized alteration (e.g., dolomitization of limestone) or the presence of thin, magnetic cap rock.

5.3.1.4 Limitations

- **Indirect Detection:** Magnetic surveys do not directly detect hydrocarbons or their host rocks. They primarily map basement and igneous features.
- **Resolution:** Lower resolution than seismic methods, making it unsuitable for detailed reservoir characterization.
- **Ambiguity:** Different geological sources can produce similar magnetic anomalies, leading to non-uniqueness in interpretation. Requires integration with other data.

5.3.2 Gravity Method

The **gravity method** in petroleum exploration measures minute spatial variations in the Earth's gravitational field caused by differences in the density of subsurface rocks. The gravitational attraction of a rock body is directly proportional to its mass, so denser rocks exert a stronger gravitational pull than less dense rocks.

5.3.2.1 Principle

The Earth's gravitational field is broadly controlled by its mass distribution. Local variations in rock density in the upper crust perturb this field, creating measurable anomalies:

- **Positive Gravity Anomaly (Gravity High):** Occurs over areas where the subsurface rock is denser than its surroundings. Examples include dense igneous/metamorphic basement rocks or compact, dense sedimentary layers.
- **Negative Gravity Anomaly (Gravity Low):** Occurs over areas where the subsurface rock is less dense than its surroundings. The most prominent example in petroleum exploration is a **salt dome**, which is significantly less dense than the surrounding sedimentary rocks.
- **Density Contrasts:** Petroleum systems are often characterized by density contrasts between various rock types (e.g., porous sands are less dense than shales; salt is much less dense than surrounding sediments; metamorphic basement is denser than overlying sediments).

5.3.2.2 Data Acquisition

Gravity surveys are conducted using highly sensitive gravimeters, which measure relative differences in gravity between measurement stations.

- **Airborne Gravity (Airborne Gravimetry and Gravity Gradiometry):**
  - **Methodology:** Utilizes specialized gyrostabilized platforms equipped with highly sensitive accelerometers in fixed-wing aircraft. Modern systems often include **gravity gradiometers**, which measure the rate of change of the gravitational field (gravitational tensor components).
  - **Instrumentation:** Airborne gravimeters can measure the full gravity vector or its derivatives. Gravity gradiometers offer higher resolution and are less susceptible to platform noise, making them excellent for resolving smaller density contrasts.

- **Advantages:** Rapid regional coverage, similar to magnetics, and good for identifying large-scale density structures. Gravity gradiometry significantly improves resolution over traditional airborne gravity.

- **Marine Gravity:**
  - **Methodology:** Gravimeters are installed on ships (hull-mounted or on specialized towed vehicles).
  - **Advantages:** Efficient for large offshore areas.
- **Land Gravity:**
  - **Methodology:** High-precision relative gravimeters are used to take measurements at discrete, precisely surveyed stations (requiring accurate GPS coordinates, especially elevation, as gravity varies significantly with altitude).
  - **Advantages:** Provides the highest resolution and most accurate data for specific onshore targets, but is slow and labor-intensive.

5.3.2.3 Applications in Petroleum Exploration

The gravity method is particularly powerful for mapping large-scale density structures, especially those involving salt.

- **Salt Dome Mapping:** This is arguably the most significant application of gravity in petroleum exploration. Salt (halite) has a typical density of 2.0-2.2 g/cm^3, whereas the surrounding shales and sandstones compact to densities of 2.3-2.6 g/cm^3 or higher at depth. This significant negative density contrast creates prominent negative gravity anomalies (gravity lows) directly over salt diapirs and domes (Nettleton, 1940). Gravity surveys are highly effective for:
  - **Detecting Salt Bodies:** Identifying buried salt domes or sheets, which are critical for forming structural traps (e.g., in the Gulf of Mexico, offshore Brazil).
  - **Delineating Salt Geometry:** Mapping the flanks and overhangs of salt bodies, which are often targets for hydrocarbon accumulation.
  - **Constraining Seismic Interpretation:** Providing independent density information that aids in building velocity models for sub-salt seismic imaging (which is extremely challenging).

- **Basin Delineation and Mapping Basement Architecture:**
    - **Sedimentary Thickness:** Thick sedimentary basins (often lower bulk density than underlying crystalline basement) generate regional gravity lows, helping to define basin extent and areas of maximum sediment accumulation.
    - **Basement Faults and Uplifts:** Large-scale variations in basement topography (e.g., horsts and grabens) that are often associated with density changes can be mapped, providing insights into basin-forming tectonics and potential structural highs.
- **Identification of Igneous Bodies:** Dense igneous intrusions (e.g., mafic sills and dykes) can generate localized positive gravity anomalies (gravity highs), helping to map their extent and impact on the petroleum system.
- **Overpressure Detection (Indirectly):** In some geological settings, highly overpressured zones (where pore fluids bear more load than the overburden) can lead to lower effective rock density, potentially creating subtle gravity lows that might indicate zones of high fluid content or drilling hazards.

5.3.2.4 Limitations

- **Lower Resolution:** Gravity surveys provide a lower resolution image compared to seismic, making them unsuitable for detailed reservoir characterization.
- **Non-Uniqueness:** Different subsurface density distributions can produce the same gravity anomaly (the fundamental problem of non-uniqueness). Interpretation requires strong geological constraints and integration with other geophysical data.
- **Depth Ambiguity:** It is often difficult to precisely determine the depth of a density anomaly from gravity data alone.
- **Topographic Sensitivity:** Gravity measurements are highly sensitive to surface topography, requiring meticulous terrain corrections, especially in rugged areas.

5.3.3 Integrated Role of Potential Field Methods

In modern petroleum exploration, magnetic and gravity methods are rarely used in isolation. Their greatest value lies in their ability to provide a cost-effective regional overview that significantly enhances and guides subsequent seismic efforts.

- **Regional Screening:** They act as the first-pass filter, identifying prospective basins or sub-basins with appropriate sedimentary thickness and structural styles.
- **Seismic Program Design:** The insights gained from potential field data help in optimizing seismic survey design, allowing geophysicists to focus seismic acquisition on areas with the highest potential and avoid areas with unprospective basement configurations or large igneous bodies.
- **Constraining Seismic Interpretation:** Gravity and magnetic data can provide independent constraints for seismic velocity models (especially for salt bodies) and help differentiate between geological features that might have similar seismic expressions but different density or magnetic signatures. For example, a seismic reflection might be caused by a volcanic sill or a gas-filled sand; magnetic data could help differentiate by confirming the presence of a magnetic igneous body.
- **Reduced Exploration Risk:** By providing a broader geological context, potential field methods contribute to a more holistic understanding of the petroleum system, ultimately reducing the overall exploration risk and optimizing resource allocation.

In conclusion, while seismic is the undisputed king of detailed imaging in petroleum exploration, magnetic and gravity methods serve as crucial scout tools. They efficiently map the regional framework, delineate the subsurface architecture of sedimentary basins, and pinpoint large-scale density or magnetic anomalies (like salt domes or basement highs), providing the essential context for more focused and expensive seismic investigations.

5.4 Borehole Geophysics (Well Logging) in Petroleum Exploration

Borehole geophysics, commonly known **as well logging or downhole geophysics, represents** a critical bridge between surface geophysical surveys and the direct geological information obtained from drill cores or cuttings. It involves lowering specialized measuring devices (probes or "sondes") into a drilled wellbore to continuously record various physical properties of the intersected rock formations and fluids. While surface methods provide broad-scale images, well logging delivers high-resolution, in-situ measurements directly within the rock volume, offering invaluable insights for petroleum geoscientists and engineers.

5.4.1 Principle

The fundamental principle of well logging is to acquire a continuous, objective profile of subsurface properties as a logging probe is traversed along the length of a borehole. Each logging tool is designed to measure a specific physical response or property of the rock and its pore fluids. These measurements are recorded at precise depth intervals, creating a "log" or "chart" that plots the measured property against depth.

The measurements reflect a range of physical characteristics, including:

- **Electrical Properties:** Electrical resistivity, conductivity, spontaneous potential. These are highly sensitive to fluid type (hydrocarbons are resistive, saline water is conductive) and porosity.
- **Radioactive Properties:** Natural gamma ray emission (from K, U, Th), attenuation of gamma rays (for density), and interaction of neutrons with hydrogen atoms (for porosity).
- **Acoustic/Elastic Properties:** Travel time of acoustic waves, indicating rock hardness, porosity, and fluid type.
- **Magnetic Properties:** Magnetic susceptibility.
- **Geometric Properties:** Borehole diameter, deviation from vertical.

By analyzing these continuous profiles, geoscientists can:

- **Identify Lithology:** Differentiate between different rock types (e.g., sandstones, shales, limestones, evaporites).
- **Quantify Petrophysical Properties:** Estimate reservoir parameters like porosity, permeability, and water saturation.
- **Detect Hydrocarbons:** Identify zones containing oil or natural gas.
- **Delineate Stratigraphic and Structural Features:** Pinpoint bed boundaries, unconformities, faults, and fractures.
- **Assess Mechanical Properties:** Provide data for wellbore stability analysis and hydraulic fracturing design.

5.4.2 Instrumentation

A well logging system comprises three main components:

1. **Logging Probe (Sonde):** This is the downhole instrument, typically a cylindrical tool designed to withstand high temperatures and pressures. Modern probes are often modular,

allowing multiple sensors to be integrated into a single "logging string" to acquire several logs simultaneously, enhancing efficiency.

2. **Logging Cable:** A multi-conductor steel-armored cable that serves several purposes:
    - **Mechanical Support:** Lowers and raises the probe within the wellbore.
    - **Power Transmission:** Supplies electrical power to the downhole instruments.
    - **Data Transmission:** Sends raw data from the probe to the surface acquisition unit in real-time.
3. **Surface Acquisition Unit (Logging Unit):** A truck-mounted or skid-mounted control center containing:
    - **Winch System:** Precisely controls the depth and speed of the logging probe.
    - **Computer System:** Manages data acquisition, processes raw signals, displays logs in real-time, and stores the final data.

A diverse array of logging tools exists, each engineered to measure specific properties:

- **Gamma Ray (GR) Log:**
    - **Principle:** Measures the natural radioactivity emitted by formations. Radioactive isotopes (primarily K-40, U-238 series, Th-232 series) are concentrated in clays and shales, while clean sandstones and carbonates are typically low in radioactivity.
    - **Application:** Primary log for lithology identification (distinguishing shales from sands/carbonates), stratigraphic correlation between wells, and identifying radioactive minerals (e.g., uranium-rich shales).
- **Resistivity Logs (Laterolog, Induction Log):**
    - **Principle:** Measure the electrical resistivity (or its inverse, conductivity) of formations. The resistivity of a porous rock is highly dependent on the resistivity of the pore fluid and the rock's porosity and tortuosity. Hydrocarbons are electrical insulators (very high resistivity), while saline formation water is highly conductive (low resistivity).
    - **Application: The most crucial logs for identifying hydrocarbon-bearing zones.** High resistivity often indicates hydrocarbons, while low resistivity suggests water saturation. Used to calculate water saturation and identify permeable zones. Different tools are used for different borehole fluid types (e.g., laterologs for conductive muds, induction logs for resistive muds).

- **Porosity Logs (Density, Neutron, Sonic):** These logs are designed to quantify the porosity of reservoir rocks, which is critical for estimating hydrocarbon volume in place. They often work together for more accurate porosity estimates.
  - **Density Log (Gamma-Gamma Log):**
    - **Principle:** Emits gamma rays into the formation and measures the number of scattered gamma rays returning to detectors. The attenuation of gamma rays is inversely proportional to the formation's bulk density.
    - **Application:** Determines bulk density, which can be used to calculate porosity (assuming known matrix density and fluid density). Also sensitive to rock type (e.g., shales are denser than porous sands) and can be used for correlation.
  - **Neutron Log:**
    - **Principle:** Emits high-energy neutrons into the formation and measures the rate at which they lose energy (slow down) or are captured. Neutrons primarily lose energy by colliding with hydrogen atoms.
    - **Application:** Measures hydrogen index, which is directly related to the volume of pore fluids (water and hydrocarbons). Used to calculate porosity, especially in shaly formations, and can help differentiate gas (lower hydrogen index than oil or water) from liquids.
  - **Sonic Log (Acoustic Log):**
    - **Principle:** Emits an acoustic (P-wave) pulse and measures the time it takes for the wave to travel through a specific length of the formation. The reciprocal of travel time is compressional wave velocity (Vp).
    - **Application:** Used to calculate porosity, estimate rock mechanical properties (important for drilling and completion), detect fractured zones (where velocity is reduced), and generate **synthetic seismograms** for seismic-to-well tie.
- **Spontaneous Potential (SP) Log:**
  - **Principle:** Measures naturally occurring electrical potentials generated by electrochemical activity between the borehole drilling fluid and formation waters across permeable bed boundaries.

 - **Application:** Primarily used to identify permeable beds (e.g., sands) within shales and to estimate the salinity of formation water.

- **Magnetic Susceptibility Log:**
  - **Principle:** Measures the magnetic susceptibility of the formation (how easily it can be magnetized).
  - **Application:** Provides a detailed magnetic profile for correlation with surface magnetic surveys and identifying magnetic minerals (e.g., magnetite, pyrrhotite) that can indicate specific lithologies (e.g., igneous intrusions) or alteration.
- **Caliper Log:**
  - **Principle:** Uses mechanical arms to measure the diameter of the borehole.
  - **Application:** Crucial for quality control (correcting other logs for borehole irregularities), identifying washouts (enlarged sections) or caving (collapsed sections), and assessing wellbore stability.
- **Imaging Logs (Optical and Acoustic Televiewers, Formation MicroImager - FMI):**
  - **Principle:** These tools provide high-resolution, oriented images of the borehole wall. Optical televiewers (in clear fluid) use cameras, while acoustic televiewers (in any fluid) use rotating ultrasonic transducers. FMI tools use arrays of micro-resistivity buttons.
  - **Application:** Provide detailed structural and sedimentological information. Used to:
    - Identify and orient fractures and faults.
    - Determine dip and strike of bedding planes and other geological features.
    - Analyze sedimentary structures (e.g., cross-bedding, laminations).
    - Detect small-scale features like vugs or dissolution features in carbonates.
    - Assess wellbore breakouts and stress orientations.
- **Directional/Deviation Logs:**
  - **Principle:** Incorporate magnetometers and accelerometers to measure the borehole's deviation from vertical (inclination) and its azimuth.
  - **Application:** Provides the precise 3D trajectory of the wellbore, essential for accurate depth positioning, geological mapping, and understanding the true spatial relationship between downhole measurements and surface locations.

- **Logging While Drilling (LWD):**
  - **Principle:** Integration of logging sensors directly into the drill collar assembly. Measurements are taken and transmitted to the surface in real-time (or near real-time) while drilling.
  - **Application:** Crucial for geosteering, where the drilling trajectory of horizontal or highly deviated wells can be adjusted in real-time to keep the wellbore within the most productive part of the reservoir, maximizing hydrocarbon recovery. Also provides early formation evaluation for making immediate drilling decisions.

5.4.3 Applications in Petroleum Exploration and Production

Borehole geophysics provides indispensable value at virtually every stage of the E&P lifecycle:

- **Formation Evaluation (The Primary Role):**
  - **Quantitative Petrophysical Properties:** The most critical application is the quantitative estimation of key reservoir parameters:
    - **Porosity (ϕ):** The fraction of rock volume composed of pore spaces, directly related to fluid storage capacity.
    - **Water Saturation (Sw):** The fraction of pore volume occupied by water. Hydrocarbon saturation (Sh) is 1−Sw. This is the ultimate determinant of whether a zone is economically productive.
    - **Lithology:** Identifying the rock type and distinguishing reservoir (e.g., sandstone, limestone) from non-reservoir (e.g., shale, salt) units.
  - **Net Pay Determination:** Accurately identifying the hydrocarbon-bearing intervals (net pay) within a reservoir, which is crucial for resource estimation.
  - **Permeability Estimation:** While not directly measured by most logs, permeability can be estimated from porosity, lithology, and other log responses.
- **Geological and Stratigraphic Correlation:**
  - **Well-to-Well Correlation:** Logs provide continuous, objective "fingerprints" of formations, allowing geoscientists to correlate lithological units, sequence boundaries, and reservoir zones between multiple wells, even over long distances. This builds the fundamental 3D stratigraphic framework of a basin.

  - **Facies Analysis:** Log patterns can indicate changes in depositional environments (e.g., fluvial channels, marine shales, carbonate platforms).
- **Well-to-Seismic Tie:**
  - **Synthetic Seismograms:** Sonic and density logs from a well are used to generate a "synthetic seismogram," which mimics a seismic trace. By comparing the synthetic seismogram to the actual seismic data at the well location, geophysicists can confidently tie seismic reflections to specific geological markers (e.g., top of reservoir, fluid contacts) identified in the well (Sheriff and Geldart, 1995).
  - **Depth Conversion:** Accurate well-to-seismic ties are essential for converting seismic time sections into true depth sections, providing precise depth estimates for reservoir intervals and structural features, which is vital for drilling planning.
  - **Seismic Calibration:** Well logs provide ground truth for calibrating seismic inversions and validating seismic attribute interpretations.
- **Geotechnical and Drilling Applications:**
  - **Wellbore Stability:** Density and sonic logs provide critical rock mechanical properties (e.g., Young's Modulus, Poisson's Ratio) that are used to predict wellbore stability and design appropriate drilling mud weights and casing programs. Caliper and imaging logs detect breakouts.
  - **Fracture Characterization:** Imaging logs are invaluable for identifying natural fractures, determining their orientation and density, and assessing their contribution to reservoir permeability or potential drilling hazards. This is especially critical in fractured reservoirs or for hydraulic fracturing design.
  - **Casing and Cement Evaluation:** Specialized logs can be run after casing and cementing to ensure the integrity of the well construction.
- **Fluid Contact and Production Monitoring:**
  - **Fluid Contacts:** Resistivity and density logs are highly effective at identifying fluid contacts (e.g., gas-oil contact, oil-water contact, gas-water contact) within a reservoir, which are critical for defining reservoir volume and production strategy.
  - **Production Logging:** Specialized logging tools (e.g., flowmeters, fluid density tools) can be run during production to monitor the flow of fluids (oil, gas, water)

from different zones in a well, helping to optimize production and diagnose problems.

  - **Time-Lapse (4D) Logging:** Repeating logs over time to monitor changes in fluid saturation or pressure in response to production or injection, similar to 4D seismic but with higher vertical resolution in the wellbore.

5.4.4 Complementary Role to Core Logging and Surface Geophysics

Well logging is not a standalone technique but forms an integral part of an integrated geoscience workflow:

- **Complementary to Core Logging:** While core provides direct physical samples for detailed analysis (e.g., mineralogy, texture), it is expensive to acquire continuously and may suffer from core loss in highly fractured or unconsolidated zones. Logs provide a continuous, in-situ record, filling in gaps where core is absent or limited.
- **Ground Truth for Surface Geophysics:** Well logs provide the essential link between the indirect measurements of surface geophysical surveys (e.g., seismic reflections, gravity anomalies) and the actual rock properties. They calibrate seismic attributes, validate inversion results, and help resolve ambiguities in surface data. Without well control, surface geophysical interpretations would be far less reliable.

In essence, borehole geophysics acts as the ultimate ground truth provider in petroleum exploration. It transforms discrete drill hole information into continuous, quantitative, high-resolution subsurface profiles, dramatically improving the accuracy and confidence of geological models, hydrocarbon resource estimates, and critical engineering decisions throughout the entire petroleum E&P lifecycle.

5.5 Advanced Geophysical Techniques and Future Trends

The petroleum industry's relentless pursuit of new hydrocarbon resources in increasingly challenging geological settings and its drive to maximize recovery from mature fields have been the primary catalysts for continuous innovation in applied geophysics. This section explores advanced techniques that are pushing the boundaries of subsurface imaging and characterization, alongside the future trends driven by technological advancements and the global energy transition.

5.5.1 Quantitative Interpretation (QI) and Reservoir Characterization

Moving beyond simply imaging geological structures, Quantitative Interpretation (QI) aims to extract direct physical properties of the reservoir (e.g., porosity, fluid saturation, lithology, permeability indicators) from seismic data. This transformation from geological image to quantitative reservoir parameter is critical for de-risking drilling and optimizing field development.

- **Seismic Inversion (Elastic Inversion):**
    - **Principle:** While traditional seismic processing focuses on generating an image of reflection coefficients, seismic inversion transforms these coefficients back into fundamental rock properties – primarily acoustic impedance (Z), and often shear impedance (Zs) and density (ρ). This is achieved by solving an inverse problem, where the goal is to find the earth model that best explains the observed seismic data (Russell, 1988).
    - **Methodology:**
        - **Acoustic Inversion:** Typically inverts for P-wave acoustic impedance (Zp =ρVp). This is often the first step and provides a direct measure of reservoir quality.
        - **Elastic Inversion (AVO Inversion):** More advanced, it uses the variation of seismic amplitude with offset (AVO) to simultaneously invert for P-impedance (Zp), S-impedance (Zs=ρVs), and often density (ρ) or the Vp/Vs ratio. These elastic properties are highly sensitive to fluid content and lithology, allowing for more robust fluid discrimination and reservoir characterization than acoustic impedance alone (Russell, 1988).
    - **Application:** Provides detailed 3D volumes of petrophysical properties that can be directly used in reservoir models, reducing uncertainty in hydrocarbon volume estimates, guiding infill drilling, and identifying sweet spots.
- **Rock Physics:**
    - **Principle:** Rock physics establishes the quantitative link between geological and petrophysical properties (e.g., lithology, porosity, fluid type, effective stress, temperature) and their corresponding seismic elastic properties (P-wave velocity, S-wave velocity, density, attenuation). It involves empirical and theoretical models that describe how rock framework and fluid properties interact (Mavko et al., 2009).

- **Application:**
  - **Predicting Seismic Response:** Using rock physics models, geophysicists can predict how a reservoir with specific porosity and fluid content (e.g., oil vs. gas vs. water) would appear on seismic data (e.g., its AVO signature). This is crucial for pre-drill risk assessment and DHI validation.
  - **Interpreting Inversion Results:** Rock physics models provide the framework to translate inverted elastic properties (e.g., Zp, Zs, Vp/Vs) into meaningful reservoir parameters like porosity, clay content, and fluid saturation.
  - **Fluid Substitution Modeling:** Simulating the seismic response of a reservoir if its pore fluids were changed (e.g., from water to oil or gas), aiding in prospect evaluation and 4D seismic interpretation.

- **Joint Inversion:**
  - **Principle:** The simultaneous inversion of multiple geophysical datasets (e.g., seismic, gravity, magnetics, CSEM, well logs) to produce a single, self-consistent subsurface model of physical properties. This approach leverages the complementary sensitivity of different methods, reducing ambiguity and improving resolution beyond what any single method could achieve (e.g., Bouzidi and Maillard, 2011).
  - **Application:** Particularly useful in complex geological settings (e.g., sub-salt, sub-basalt) where one method might struggle to provide a complete image. Combining seismic (high resolution) with gravity (sensitive to density contrasts) can greatly improve the accuracy of subsurface density models.

5.5.2 Microseismic Monitoring (for Unconventional Resources)

With the rise of unconventional resources (e.g., shale gas, tight oil), microseismic monitoring has emerged as a critical technology for optimizing hydraulic fracturing operations.

- **Principle:** Hydraulic fracturing involves injecting high-pressure fluid into a wellbore to create and propagate artificial fractures in low-permeability reservoir rocks. These fracturing events, as well as associated stress relaxation, generate tiny seismic signals (microseisms), typically with magnitudes below zero. A network of highly sensitive geophones (or DAS fibers) deployed in nearby monitoring wells or on the surface detects these microseisms.

- **Methodology:**
    - **Passive Listening:** Unlike active seismic, microseismic monitoring is a passive technique, simply "listening" to the Earth's natural or induced rumblings.
    - **Source Location:** By precisely timing the arrival of microseismic waves at multiple sensors, the location (hypocenter), time, and magnitude of each fracturing event can be determined.
- **Application:**
    - **Mapping Fracture Networks:** Creating a real-time 3D map of the stimulated reservoir volume (SRV) and the geometry (length, height, orientation) of induced fractures (Rutledge et al., 2004). This is paramount for understanding how effectively the reservoir is being stimulated.
    - **Optimizing Frac Design:** Providing feedback to adjust injection rates, fluid volumes, and proppant concentrations to maximize stimulation efficiency.
    - **Well Spacing and Placement:** Guiding the optimal spacing of future wells based on the observed fracture distribution.
    - **Interference Detection:** Identifying potential fracture hits between adjacent wells.
    - **Fault Activation Monitoring:** Detecting whether fracturing has reactivated pre-existing faults, which is important for understanding potential induced seismicity and fluid containment.

5.5.3 Electromagnetic (EM) Methods (Emerging Applications)

While EM methods have been used in mining for decades, their application in deep offshore petroleum exploration is a more recent development, primarily focused on **Marine Controlled-Source Electromagnetic (CSEM)**.

- **Principle:** Hydrocarbons (oil and gas) are electrical insulators, meaning they have very high electrical resistivity. Saline formation water, on the other hand, is highly conductive. CSEM exploits this fundamental contrast. A powerful electrical dipole transmitter is towed near the seafloor, emitting a controlled electromagnetic field. This field induces electric currents in the subsurface. Receivers (electric field and/or magnetic field sensors) deployed on the seafloor record the resulting EM field.

- **Methodology:**
    - The EM signal propagates through the seawater and the seafloor. When it encounters a highly resistive hydrocarbon-filled reservoir, the EM energy is channeled through this resistive layer, creating a measurable anomaly at the receivers (Edwards, 2005).
    - The depth of penetration and sensitivity are controlled by the frequency of the transmitted EM energy.
- **Application:**
    - **Direct Hydrocarbon Detection (DHI):** CSEM can directly detect highly resistive hydrocarbon-filled reservoirs, especially in deepwater environments where drilling costs are immense. It helps to differentiate between resistive pay zones and conductive water-bearing sands, reducing the risk of drilling dry wells.
    - **Fluid Contact Mapping:** Can help delineate fluid contacts within a reservoir.
    - **4D CSEM (Time-Lapse CSEM):** Repeating CSEM surveys over time can monitor changes in reservoir resistivity due to fluid movement (e.g., water replacing hydrocarbons during production), providing valuable insights for reservoir management.
- **Advantages:** Provides an independent geophysical measurement that directly responds to fluid type, complementing seismic data and reducing drilling risk.
- **Limitations:** Lower resolution than seismic, sensitive to conductive overburden, and challenges in complex geological settings (e.g., basalt or carbonates).

5.5.4 Machine Learning and Artificial Intelligence (AI)

The explosion of data volume, the increasing complexity of geological models, and the need for faster decision-making have positioned Machine Learning (ML) and Artificial Intelligence (AI) as transformative forces in petroleum geophysics.

- **Automated Interpretation and Feature Extraction:**

    - **Seismic Facies Classification:** Training ML algorithms to automatically classify seismic facies (patterns of seismic reflections that relate to depositional environments or lithologies) across large 3D seismic volumes.
    - **Fault Detection and Tracking:** AI models can be trained to recognize and automatically map faults, fractures, and other structural discontinuities from seismic data, accelerating structural interpretation and reducing human interpretation time.
    - **Anomaly Detection:** Quickly identifying subtle or anomalous features (e.g., DHIs, complex stratigraphic features) that might be overlooked by human interpreters.
- **Predictive Modeling:**
    - **Reservoir Property Prediction:** ML models can be trained on well log data (known porosity, permeability, fluid saturation) and corresponding seismic attributes to predict these reservoir properties in undrilled areas of a seismic volume.
    - **Drilling Optimization:** Predicting drilling hazards (e.g., overpressure zones, unstable formations) from seismic and well log data.
- **Big Data Analytics:** AI/ML frameworks are essential for managing, processing, and extracting meaningful insights from the petabytes of data generated by modern geophysical surveys (e.g., ultra-large 3D/4D seismic, high-density OBS/OBN).
- **Inverse Problem Solving:** Applying ML/AI to accelerate and improve geophysical inversion algorithms, particularly for complex problems like Full Waveform Inversion.

5.5.5 New Acquisition Technologies

Continuous innovation in sensor technology and deployment platforms is driving the next generation of geophysical data acquisition.

- **Broadband Seismic:**
    - **Principle:** Aims to acquire seismic data with a much wider range of frequencies than conventional seismic (both low and high frequencies). Low frequencies improve penetration and imaging below complex overburden, while high frequencies enhance resolution.

  - **Methodology:** Achieved through specialized marine source arrays (e.g., variable airgun arrays) and receiver designs, as well as advanced processing techniques (e.g., deghosting).
  - **Application:** Provides sharper images, better resolution of thin beds, and improved penetration in challenging geological settings.
- **Node Technology (Ocean Bottom Nodes - OBN):**
  - **Principle:** Autonomous, cable-free seismic receivers containing multi-component geophones and hydrophones. They are deployed on the seafloor and later retrieved.
  - **Advantages:** Offers superior flexibility in survey design (can be deployed around obstacles), provides full-azimuth and long-offset coverage (crucial for imaging complex structures), and captures a full wavefield (P-waves and S-waves), leading to higher quality and more comprehensive data compared to towed streamers (Sørensen et al., 2018).
  - **Application:** Revolutionizing imaging in complex deepwater environments, congested fields with existing infrastructure, and for 4D monitoring.
- **Distributed Acoustic Sensing (DAS):**
  - **Principle:** As discussed in previous sections, DAS transforms a standard fiber optic cable into a continuous array of thousands of seismic sensors. The system detects subtle changes in the light backscatter along the fiber, which are caused by strain induced by acoustic waves (Parker et al., 2014).
  - **Application:**
    - **Vertical Seismic Profiles (VSP):** High-density VSP data from wells, providing superior resolution near the wellbore.
    - **Microseismic Monitoring:** Continuous, high-density monitoring of hydraulic fracturing.
    - **Reservoir Surveillance:** Monitoring flow within the wellbore, reservoir changes, and even integrity of well completions.
    - **Permanent Monitoring:** Fiber optic cables can be permanently installed in wells or on the seafloor for continuous, long-term 4D monitoring.

5.5.6 Geophysics for Carbon Capture, Utilization, and Storage (CCUS)

As the world transitions to a lower-carbon economy, geophysical methods are being repurposed and adapted to address the critical challenge of climate change, particularly in CCUS.

- **Objective:** To safely and permanently store large volumes of captured CO2 in deep geological formations (e.g., depleted oil/gas reservoirs, deep saline aquifers). Geophysics is essential for:
    - **Site Characterization:** Identifying suitable subsurface storage sites with adequate capacity, injectivity (ability to inject CO2), and, crucially, robust seal integrity to prevent CO2 leakage. This involves detailed seismic imaging, well logging, and rock physics analysis.
    - **Injection Monitoring:** Using **4D seismic** (time-lapse seismic) to track the injected CO2 plume over time. Changes in seismic velocity and amplitude due to the presence of CO2 (which is less dense and more compressible than water) are measured, allowing geophysicists to monitor the CO2's movement, ensure it stays within the intended storage complex, and detect any potential leakage pathways (Lumley, 2010).
    - **Geomechanical Monitoring:** Assessing and monitoring changes in subsurface stress and pore pressure due to CO2 injection, using microseismic monitoring to detect induced seismicity.
- **Application to Geothermal Exploration:** Geophysical methods (seismic, EM, gravity, magnetics) are also increasingly vital for identifying and characterizing geothermal reservoirs, which rely on permeable, hot rock volumes.

5.5.7 Challenges and Opportunities

Despite these rapid advancements, petroleum geophysics faces ongoing challenges that also represent significant opportunities for future innovation:

- **Subsurface Imaging in Complex Environments:** Imaging beneath highly attenuating or scattering media (e.g., thick basalt flows, complex salt bodies, highly fractured zones) remains a major challenge. Further development of FWI, new acquisition geometries, and advanced processing algorithms are needed.

- **Data Volume and Computational Demands:** The sheer volume of data generated by modern surveys (e.g., ultra-large 3D/4D, high-density DAS) requires continuous advancements in high-performance computing (HPC), cloud-based processing, and efficient data management strategies.
- **Resolution vs. Depth:** The fundamental trade-off between resolution and depth of penetration persists. Pushing for higher resolution at greater depths requires more powerful sources and more sensitive receivers.
- **Reducing Uncertainty and Ambiguity:** While multi-physics integration and quantitative interpretation reduce ambiguity, the inherent non-uniqueness of geophysical inverse problems remains. Further integration with geological and petrophysical constraints, and advanced joint inversion techniques, are key.
- **From Data to Decisions:** The challenge is to bridge the gap between complex geophysical data and actionable business decisions. AI and ML are crucial in automating parts of the interpretation workflow and prioritizing targets.
- **Energy Transition Integration:** Adapting existing geophysical tools and developing new ones for applications in new energy domains (CCUS, geothermal, hydrogen storage, critical minerals) is a major opportunity and necessity.
- **Talent Development:** The need for a new generation of geophysicists skilled in data science, advanced modeling, and integrated multi-physics interpretation, coupled with strong geological fundamentals.

In conclusion, petroleum geophysics is a vibrant and technologically advanced field, constantly pushing the boundaries of what is possible in subsurface imaging and characterization. The ongoing development of sophisticated processing algorithms, quantitative interpretation techniques, and novel acquisition platforms, coupled with the transformative potential of AI and ML, will continue to drive more efficient, lower-risk, and environmentally responsible hydrocarbon exploration and will play an increasingly critical role in the broader energy transition.

5.6 Challenges and Future Directions

The field of applied geophysics in petroleum exploration has achieved remarkable feats over the past decades, transforming from a nascent science into a sophisticated, indispensable discipline.

However, the industry continually faces evolving challenges, driven by the increasing complexity of geological targets, economic pressures, technological advancements, and the global energy transition. These challenges simultaneously open up exciting avenues for future research, development, and application.

5.6.1 Subsurface Imaging in Complex Environments

One of the most persistent and formidable challenges in petroleum geophysics is accurately imaging the subsurface in highly complex geological settings. Many of the easily accessible, structurally simple hydrocarbon traps have already been discovered, forcing exploration into areas with increasingly difficult imaging conditions.

- **Complex Overburden:**
    - **Salt Bodies:** Regions like the Gulf of Mexico, Brazil's Pre-Salt, and West Africa are characterized by massive, often irregularly shaped salt bodies. Salt has very high seismic velocity and tends to scatter and absorb seismic energy, creating significant imaging challenges beneath it. Accurate sub-salt imaging requires sophisticated acquisition (e.g., wide-azimuth, long-offset, OBN) and advanced processing techniques like Full Waveform Inversion (FWI) to build precise velocity models (Virieux and Operto, 2009). The ongoing development of robust FWI algorithms capable of handling very complex velocity fields is critical.
    - **Basalt/Volcanic Flows:** Thick sequences of basaltic lavas or volcanic sills and dykes (e.g., offshore Brazil, West Africa, and parts of the North Atlantic) pose similar, if not greater, challenges than salt. Basalt is highly attenuative, highly scattering, and often extremely heterogeneous, making it difficult for seismic waves to penetrate and return usable reflections from below. Research into new source technologies, low-frequency seismic, and non-seismic methods (e.g., CSEM) to aid in sub-basalt imaging is paramount.
    - **Highly Fractured or Karstified Carbonates:** These reservoirs can be highly heterogeneous, causing significant seismic scattering and anisotropy. Imaging and characterizing these features requires full-azimuth seismic data and advanced processing that accounts for anisotropy.

- **Deepwater and Ultra-Deepwater Exploration:** As exploration moves into deeper waters and greater depths, the challenges intensify. Longer seismic travel paths lead to greater signal attenuation and lower resolution. This necessitates more powerful and broadband seismic sources, longer streamers/offsets, and ocean bottom node (OBN) acquisition for superior signal-to-noise ratio and full-azimuth coverage.
- **Shallow Hazards:** While the focus is often on deep targets, imaging and identifying shallow drilling hazards (e.g., shallow gas pockets, gas hydrates, weak formations, active faults) is crucial for drilling safety and efficiency. High-resolution shallow seismic, often with specialized sources and receivers, is an ongoing area of development.

5.6.2 Data Volume and Computational Demands

Modern geophysical surveys, particularly large 3D and 4D seismic acquisitions, generate immense volumes of data – often petabytes for a single project. The processing and interpretation of this "big data" pose significant computational challenges.

- **High-Performance Computing (HPC):** The move from 2D to 3D and now 4D seismic, coupled with computationally intensive algorithms like FWI and advanced migration, has pushed the boundaries of conventional computing. There is a continuous demand for greater processing power, parallel computing architectures, and efficient data storage solutions.
- **Cloud Computing:** The scalability and flexibility of cloud computing platforms are becoming increasingly attractive for seismic processing. Cloud solutions allow companies to access vast computing resources on demand, reducing capital expenditure on in-house supercomputers and enabling faster turnaround times for large projects.
- **Data Management and Workflow Integration:** Effectively managing, cataloging, and integrating diverse geophysical datasets (seismic, well logs, gravity, magnetics, CSEM) from various vintages and formats is a major challenge. Seamless data workflows, robust data governance, and common data standards are essential to maximize data value and minimize errors.

5.6.3 Reduced Cycle Time and Automation

The petroleum industry operates in a dynamic and often volatile market. There is constant pressure to accelerate the exploration-to-production cycle to reduce costs and bring resources to market faster.

- **Faster Acquisition:** Development of more efficient acquisition platforms (e.g., more streamers per vessel, multi-source arrays, autonomous nodes, uncrewed surface vessels for seismic) and faster deployment/retrieval systems.
- **Real-Time Processing and Interpretation:** The goal is to move towards near real-time processing and interpretation, allowing for immediate feedback during drilling operations (e.g., LWD for geosteering) or dynamic adjustments during acquisition. This is where AI and ML play a transformative role.
- **Automation:** Automating repetitive tasks in data processing, quality control, and even initial interpretation (e.g., automated fault picking, horizon tracking, facies classification using ML) frees up geoscientists to focus on higher-level geological interpretation and decision-making.

5.6.4 Resolution and Ambiguity

Despite advancements, improving the resolution of subsurface images and reducing the inherent non-uniqueness (ambiguity) in geophysical interpretations remain core challenges.

- **Increased Resolution:** The drive for higher resolution aims to delineate thinner beds, smaller faults, and more subtle stratigraphic features that can be critical for defining reservoir architecture and identifying bypassed pay. This requires broader bandwidth seismic data (lower and higher frequencies), more precise velocity models, and advanced inversion techniques.
- **Non-Uniqueness of Inverse Problems:** All geophysical methods solve inverse problems, where multiple subsurface models can produce similar observed data. This inherent ambiguity requires robust geological constraints and the integration of multiple geophysical methods (joint inversion) to arrive at the most probable and geologically realistic solution.
- **Quantitative Uncertainty Estimation:** Providing not just a best-estimate model but also an understanding of the uncertainties associated with geophysical predictions (e.g.,

uncertainty in porosity, fluid saturation) is crucial for robust risk assessment and resource quantification. Bayesian inversion and stochastic modeling are actively being developed for this purpose.

5.6.5 Energy Transition and Diversification of Applications

The global imperative to transition to a lower-carbon energy future is profoundly impacting the petroleum industry. Geophysics, as a core subsurface imaging discipline, is uniquely positioned to play a vital role in this transition.

- **Carbon Capture, Utilization, and Storage (CCUS):** This is a rapidly growing application. Geophysics is essential for:
    - **Site Characterization:** Identifying and assessing suitable deep geological formations (saline aquifers, depleted oil/gas reservoirs) for safe and long-term CO2 storage. This involves detailed seismic imaging of reservoir capacity, injectivity, and, crucially, the integrity of sealing formations to prevent CO2 leakage.
    - **Monitoring CO2 Plume: 4D seismic** is the primary tool for monitoring the injected CO2 plume over time, tracking its migration and ensuring it remains contained within the storage complex (Lumley, 2010). Any changes in seismic attributes (velocity, amplitude) can indicate CO2 presence.
    - **Geomechanical Monitoring:** Using microseismic monitoring to ensure that CO2 injection does not induce significant seismicity or compromise the integrity of the storage complex.
- **Geothermal Energy Exploration:** Geophysics (seismic, EM, gravity, magnetics) is crucial for identifying and characterizing geothermal reservoirs, which involve identifying zones of high heat flow, permeable fractures, and fluid pathways in the subsurface.
- **Subsurface Hydrogen Storage:** As hydrogen gains prominence as a clean energy carrier, the need for safe and efficient subsurface storage (e.g., in salt caverns or depleted gas fields) will grow. Geophysical monitoring will be essential for ensuring the integrity of these storage sites.
- **Critical Minerals Exploration:** Many critical minerals are associated with specific geological environments. Geophysical techniques, adapted from petroleum (e.g., deep seismic for crustal structures, EM for conductivity anomalies), are finding increasing

application in delineating these resources (e.g., lithium brines, deep-seated rare earth deposits).

- **Decommissioning and Environmental Monitoring:** Geophysics can also be used to map and monitor legacy infrastructure (e.g., abandoned wells, pipelines) and to monitor environmental impacts of past operations (e.g., groundwater contamination).

5.6.6 Machine Learning (ML) and Artificial Intelligence (AI) - The Game Changers

As previously discussed, ML and AI are not just trends but fundamental shifts in how geophysical data will be processed and interpreted.

- **Enhanced Interpretation Efficiency:** AI can automate mundane and repetitive tasks, allowing geophysicists to focus on complex, high-value interpretations.
- **Superior Anomaly Detection:** ML algorithms can identify subtle patterns and anomalies in large datasets that might be missed by human eyes.
- **Improved Predictive Capabilities:** AI can build robust predictive models for reservoir properties and drilling risks based on vast historical data.
- **Accelerated Inversion:** Using neural networks and other ML techniques to speed up complex geophysical inversion processes, enabling quicker model updates and iterations.
- **Uncertainty Quantification:** AI can assist in more robustly quantifying uncertainties in subsurface models by exploring a wider range of possible solutions.
- **Data Integration and Knowledge Extraction:** AI can effectively integrate disparate data types (geophysical, geological, geochemical, drilling) to extract deeper insights and build more comprehensive subsurface knowledge bases.

5.6.7 Talent Development and Cross-Disciplinary Integration

The evolving landscape of geophysics demands a new breed of geoscientists.

- **Multi-Disciplinary Skills:** Future geophysicists need a strong foundation in traditional geophysical principles combined with expertise in data science, programming (e.g., Python), machine learning, cloud computing, and advanced numerical modeling.
- **Cross-Domain Knowledge:** As geophysics diversifies into CCUS, geothermal, and other new energy ventures, geophysicists will need to acquire a broader understanding of these new domains.

- **Collaboration:** Increased collaboration between geophysicists, geologists, reservoir engineers, data scientists, and computational scientists will be crucial for solving complex subsurface problems.

In conclusion, the journey of applied geophysics in petroleum exploration is far from over. While the challenges of imaging increasingly complex deep targets persist, the ongoing revolution in data acquisition (e.g., OBN, DAS), processing (e.g., FWI, broadband), and interpretation (e.g., AI/ML, QI) promises to unlock new frontiers. Furthermore, the pivotal role of geophysics in the energy transition, particularly in CCUS and geothermal, ensures its continued strategic importance and guarantees a dynamic and exciting future for the discipline.

Bibliography

- Kearey, Philip, and Michael Brooks. An introduction to geophysical exploration. 3rd ed. Wiley-Blackwell, 2002.
- Lowrie, William. (2007) Fundamentals of Geophysics. Cambridge University Press, Cambridge, UK. https://doi.org/10.1017/CBO9780511807107.
- Telford, W.M., Geldart, L.P. and Sheriff, R.E. (1990) Applied Geophysics. 2nd Edition, Cambridge University Press, Cambridge, 770. https://doi.org/10.1017/CBO9781139167932.
- Turcotte, Donald L., and Gerald Schubert. Geodynamics. 2nd ed. Cambridge University Press, 2002.
- Sharma, Subhash C. Borehole geophysics: Exploration, reservoir characterization and monitoring. Elsevier, 2009.
- Breiner, Sheldon. (1973). Applications Manual for Portable Magnetometer.
- Goodkind, J. M. (1999). The superconducting gravimeter. *Review of Scientific Instruments, 70*(11), 4131-4152.
- IAEA. (2003). *Guidelines for radioelement mapping using gamma ray spectrometry data*. International Atomic Energy Agency.
- Keller, G. V., & Frischknecht, F. C. (1966). *Electrical methods in geophysical prospecting*. Pergamon Press.
- Kinsler, L. E., Frey, A. R., Coppens, A. B., & Sanders, J. V. (2000). *Fundamentals of acoustics* (4th ed.). John Wiley & Sons.
- Lillesand, T. M., Kiefer, R. W., & Chipman, J. W. (2015). *Remote sensing and image interpretation* (7th ed.). John Wiley & Sons.
- Nabighian, M. N. (Ed.). (1987). *Electromagnetic methods in applied geophysics: Theory* (Vol. 1). Society of Exploration Geophysicists.
- Parasnis, D. S. (1997). *Principles of applied geophysics* (5th ed.). Chapman & Hall.
- Reynolds, J. M. (2011). *An introduction to applied and environmental geophysics* (2nd ed.). John Wiley & Sons.
- Shan, J., & Toth, C. K. (Eds.). (2009). *Topographic laser ranging and scanning: principles and processing*. CRC Press.

- Sheriff, R. E., & Geldart, L. P. (1995). *Exploration seismology* (2nd ed.). Cambridge University Press.
- Telford, W. M., Geldart, L. P., Sheriff, R. E., & Keys, D. A. (1990). *Applied geophysics* (2nd ed.). Cambridge University Press.

Appendix A: Glossary of Terms

Airborne geophysics: The use of geophysical instruments mounted on aircraft to map the subsurface over large areas.

Cross-well seismic tomography: A type of seismic tomography that uses seismic waves generated by multiple sources to image the subsurface between boreholes.

Electrical resistivity tomography (ERT): A geophysical method that uses electrical currents to map the subsurface resistivity.

Electromagnetic (EM) exploration: A geophysical method that uses electromagnetic fields to map the subsurface structure and to identify potential mineral deposits or other geological features.

Gravity exploration: A geophysical method that measures the Earth's gravity field. The gravity field is affected by the density of the subsurface material.

Ground penetrating radar (GPR): A geophysical method that uses radar waves to image the subsurface to depths of several meters.

Induced polarization (IP): A geophysical method that measures the electrical polarization of the subsurface. IP is particularly useful for identifying mineral deposits, such as sulfide ores and gold deposits.

Magnetic exploration: A geophysical method that measures the Earth's magnetic field. The magnetic field is affected by the magnetization of the subsurface rocks.

Seismic tomography: A geophysical method that uses seismic waves to image the subsurface structure.

Surface seismic tomography: A type of seismic tomography that uses seismic waves generated at the surface to image the subsurface.

Borehole geophysics: The use of geophysical instruments lowered into boreholes to investigate the subsurface from within.

- **Applied geophysics:** The use of geophysical methods to solve practical problems in engineering, geology, and other fields.
- **Mineral exploration:** The search for and evaluation of potential mineral deposits.
- **Groundwater exploration:** The search for and evaluation of potential groundwater aquifers.

- **Geotechnical engineering:** The application of engineering principles to the study and use of geo-materials, such as soil and rock.
- **Environmental remediation:** The process of cleaning up environmental contamination.
- **Archaeological research:** The study of past human cultures and societies through the excavation and analysis of archaeological sites.

Appendix B: Mathematical Formulas

Here is a summary of some of the most important mathematical formulas used in applied geophysics.

Seismic Waves

The velocity of seismic waves in a homogeneous and isotropic medium is given by the following equation:

$v = \mathrm{sqrt}(K/\rho)$

where:

- v is the wave velocity
- K is the bulk modulus
- ρ is the density

The reflection coefficient for seismic waves at an interface between two media is given by the following equation:

$R = (Z2 - Z1)/(Z2 + Z1)$

where:

- R is the reflection coefficient
- Z1 is the acoustic impedance of the first medium
- Z2 is the acoustic impedance of the second medium

The acoustic impedance of a medium is given by the following equation:

$Z = \rho v$

Gravity

The gravitational acceleration due to a point mass is given by the following equation:

$g = GM/r^2$

where:

- G is the universal gravitational constant
- M is the mass of the point mass
- r is the distance from the point mass

The gravity anomaly at a point on the Earth's surface is given by the following equation:

$\delta g = G\rho\Delta V/r^2$

where:

- δg is the gravity anomaly
- ρ is the density contrast between the subsurface anomaly and the surrounding material
- ΔV is the volume of the anomaly
- r is the distance from the center of the anomaly to the point of observation

Magnetics

The magnetic field due to a magnetic dipole is given by the following equation:

$B = \mu 0m/4\pi r^3$

where:

- B is the magnetic field
- $\mu 0$ is the permeability of free space
- m is the magnetic moment of the dipole
- r is the distance from the dipole to the point of observation

The magnetic anomaly at a point on the Earth's surface is given by the following equation:

$\Delta B = \mu 0m/4\pi r^3 \cos(\theta)$

where:

- ΔB is the magnetic anomaly
- m is the magnetic moment of the subsurface anomaly
- r is the distance from the center of the anomaly to the point of observation
- θ is the angle between the magnetic field and the vertical

Electrical Resistivity

The electrical resistivity of a material is given by the following equation:

$\rho = V/I$

where:

- ρ is the electrical resistivity
- V is the voltage difference between two points in the material
- I is the current flowing through the material

The electrical conductivity of a material is the reciprocal of the electrical resistivity.

References:

- Telford, William M., L. P. Geldart, and R. E. Sheriff. Applied geophysics. Cambridge University Press, 1990.

- Kearey, Philip, and Michael Brooks. An introduction to geophysical exploration. Wiley-Blackwell, 2013.
- Sharma, Subhash C. Borehole geophysics: Exploration, reservoir characterization and monitoring. Elsevier, 2009.

Index

R

S

V

W

Y

www.ingramcontent.com/pod-product-compliance
Lightning Source LLC
LaVergne TN
LVHW060038210926
848481LV00013B/3901